"Hold on, safety profession, this one's going to rock you! Every once in awhile, something comes along that holds the promise of unlocking the future. I believe *Values-Driven Safety* is the key to that future. This is powerful stuff, an elixir for a profession searching for a 'better way.' VIRO is that 'better way!' At last, a way to elevate safety's strategic significance to the business process. At a time when answers are needed most, *Values-Driven Safety* is truly a path to a new frontier of safety success."

—Larry Hansen, Manager, Safety Management and
Organization Performance
WAUSAU INSURANCE

"The values-driven approach is leading-edge safety. The measures of progress included in this powerful book may be the next breakthrough for enlightened leaders. The insights provided can identify the values in organizational cultures needed to achieve lasting results. This valuable resource for establishing your pathway to minimizing risks and limiting losses is A WINNER!"

—Douglas Tambor, Director, Safety and Industrial Hygiene
JOHNSON & JOHNSON NORTH AMERICA

"If you are looking for a new "paradigm" in safety, look no further. *Values-Driven Safety* outlines a new process that is ongoing, self-renewing, and enlightening. This culture- and value-driven process integrates safety into the total business plan of a company. I will be sending some of my safety professionals to Don's seminars."

—Charles R. Dancer, Director, Safety and Loss Prevention
ALLIED SIGNAL

"For many years, occupational safety and health practitioners have complained of 'getting no respect' from their management. But VIRO strikes a chord that will force organizations to rethink the role and potential added value of the work of safety professionals. This is must-reading for forward-thinking organizations and may help generate some of that management respect so desperately needed by the safety profession. *Values-Driven Safety* is the future direction of the safety profession and VIRO is the ticket to get there."

—Lewis C. Booker, Manager, Loss Prevention
CHESEBROUGH-PONDS USA

"*Values-Driven Safety* is chock full of useful insights for managing organizations and life, masquerading as a bold, innovative approach to safety in the workplace. This book refuses to be ignored. It will bring the safety profession 'out of the closet' and enlighten mainstream corporate America."
—Jack McFadden, Vice President
MARKETING CORPORATION OF AMERICA

"The VIRO concept is clearly an important new way of addressing safety management. *Values-Driven Safety* makes a valuable contribution to our discipline. It's cutting-edge stuff, and I believe most instructors will want their students to master it."
—Robert E. McClay, Associate Professor, Department of Safety Sciences
INDIANA UNIVERSITY OF PENNSYLVANIA

Values-Driven Safety delivers a powerful and timely message. The principles and techniques outlined in this insightful new look at the safety function should be a prescription for rethinking how we deal with safety issues as we enter the 21st century. Human resources and line operating executives will find it equally appealing, useful, and provocative.
—Ted C. Mullins, Vice President (Retired)
UNILEVER U.S.

"The safety profession has reached the point where it must evolve to the next level. VIRO could be the evolutionary tool that moves safety into the 21st century. As the safety profession struggles with measuring performance, VIRO gives us a new and innovative approach to measurement, while integrating the safety function into the overall day-to-day operation of business."
—Kimberly K. Querrey, Manager, Corporate Safety and Health
IMCO RECYCLING

"These principles are timeless truths, whether the context is managing safety programs, the recruiting process, developing organizations, or simply leading your life. The simple and effective tools suggested here cut directly to the issues. Thoughtfully presented and easily read, this book should be in every Human Resource professional's library."
—Barry D. Jesse, Director, Human Resources
ALBANY INTERNATIONAL

"Behaviorally sound, *Values-Driven Safety* is a must for improving the quality of human interactions in the workplace. The ideas presented here have far-reaching applications for business and personal success, as well as a safe workplace."
—Michael W. Kessler, PhD, Behavioral Psychologist
COMMUNITY WORKSHOP

Values-Driven
SAFETY

Reengineering
Loss Prevention
Using

Value
Inspired
Resource
Optimization

Donald J. Eckenfelder, CSP, P.E.

Government Institutes
Rockville, Maryland

Government Institutes, Inc., 4 Research Place, Suite 200, Rockville, Maryland 20850

00 99   5   4   3   2

ISBN: 0-86587-532-4

Printed in the United States of America

This book is dedicated to my four children, eleven grandchildren (as well as two on the way and others yet to come), my daughter-in-law, and three sons-in-law. My wife, Barbara, stands on a higher plane than one who should have a book dedicated to her...at least in my eyes.

TABLE OF CONTENTS

The road to VIRO starts here . . .

The Prologue provides important signposts for your VIRO journey.

Section I: Introduction to VIRO

Section II: Suggested Values and Rationale Options and Reflections

Section III: Implementation and Measurement

The VIRO voyage ends here.

List of Figures and Tables

FOREWORD

BEEN THERE, DONE IT—not until you've read this book! They don't teach this stuff at school, seminars, or work. Don't make another plan or business decision without understanding the strategy that Don Eckenfelder outlines in this book.

Most safety and loss prevention professionals can easily apply engineering controls to reduce risk, provide training on a safety topic, or write and implement a safety procedure. But, when it comes to integrating safety into an organization's core business values and planning, we fail miserably as a profession. It's time for a change!

For the first time, someone in our profession has developed and articulated a management strategy that will change the way we currently manage and approach our profession of safety and loss prevention. This book is not for the timid or weak of heart; it will challenge you to rethink your entire approach to managing safety, loss prevention, or any other management function.

This stuff really works. I've been integrating Don's value-based management process into my organization's Safety and Loss Prevention System ever since he introduced the concept to me in 1993. It doesn't matter if we're working in Europe, North America, or Asia, the strategy works and our performance is improving every year.

Need an acid test to determine the success of a goal or project before it's launched, an early warning device to detect scud missiles, or a bolt of lightning to wake people up in your organization? Try a value-based approach to managing safety and loss prevention. The premise of this book has become a fundamental strategy that I apply to all business planning and operational tasks.

Don's strategy provides a template for any existing program, process, or system that you wish to improve. More importantly, this process is a catalyst for plant or company wide systematic changes, and it's in a language that everybody understands.

If you've ever asked yourself what drives a company to outperform its competition year after year, a sports team to dominate their field, a site work team to set performance records in a company, or an individual to exceed and drive for improvement, then you need to read this book.

Craig W. Bennett, CSP
Corporate Director of Safety and Loss Prevention
Hasbro, Inc.

PREFACE

The most powerful force in the world is an idea whose time is at hand. I believe that I have stumbled upon a very powerful concept that has the potential to elevate the way workplace safety and health is handled throughout the world and to provide side benefits that, if recognized and applied, could shed new light on the safety profession and literally change the way the world works. Churchill's commentary on man states that "Occasionally, man will stumble over the truth, but he will usually pick himself up and carry on." I have decided to stop and ponder the truth that I believe has been laid before me and to share it with others. I hope others will examine my ideas and assist me in polishing and developing them.

Prophets have—in parables and allegories—likened ideas to seeds. I borrow the seed analogy and apply it to a new way to manage risks and minimize loss. The idea needs to be planted in good soil, watered, and nurtured; then we will know if it is a good seed and if it will produce a viable plant that will reach maturity, go to seed, and propagate itself. The seed can represent the concepts in this book. The soil will be enlightened enterprises that see the virtue of limiting loss by laying a foundation of values upon which their human resources can comfortably avoid undesired outcomes. The watering will be provided by myself and others who anticipate this approach fruitful.

Historically, the safety profession has had much to offer, but has lacked the confidence and the vision to see how the application of principles they know and use could impact business and society in profound ways. We have also lacked initiative, timing, and a credible spokesperson. I hope I will see that change during my lifetime. The concept of managing outcomes by installing values has the potential to enhance every area of importance to the human family. Similar to medicine, loss prevention is a diagnostic and prescriptive activity, with results dependent upon the accuracy of the diagnosis. Safety and health practices can provide a wonderful laboratory and workshop to learn and experiment with concepts and techniques in a supportive environment, relatively free of corrupting influences.

If a child attends nine years of school and graduates from high school but cannot read and write proficiently, we are witnessing an "accident." If someone is sick, goes to a health care facility, and exits not only unhealed but with more problems than she entered with, again we see an "accident." If a person is minding his own business and is mugged or shot because he unwittingly ventured into an area that was not well policed and had a different set of survival rules than he was familiar with, that is an "accident." All these scenarios result in undesirable outcomes. We can refer to them either as accidents or undesirable outcomes, but they are the same. I suggest that the methods to avoid these wrong outcomes are known and are the same in all cases. We need to define them, understand them, and apply them to our universal benefit.

I will be an early spokesperson for these ideas, but I hope many will join me and greatly enhance my embryonic efforts.

What got me started on all this and what gives me the audacity to try writing a book on such a profound subject? I need to answer that question before I expect you to read and examine this work.

Several years ago, I noticed that the ideas being exposed concerning behavior-based safety processes embodied many of the truths that I had come to believe in during my thirty-plus year career in safety and health. In particular, I became interested in the work of Tom Krause and visited with him. He gave me a copy of his book, *The Behavior-Based Safety Process*, and I read it cover to cover and highlighted it. I found myself using the concepts in my consulting practice and came to the conclusion that the power of exposed thinking clearly transcended safety and health. In a brief telephone conversation with Tom, I asked him if he realized that his ideas could solve many of the problems that world leaders danced around but could neither bring into clear focus nor solve. He answered that he was well aware of the potential of his theories.

I have waited to see broader application of behavior-based loss prevention by Tom or other advocates, but have been disappointed with repetition and misapplications. Perhaps I have been asking too much of a valid concept that lacked the power to "move mountains."

More recently, I have heard the word "values" permeating conversations, speeches, books, and articles on a wide variety of topics. They range from families (e.g., Richard Eyre's best seller, *Teaching Your Children Values*, which was probably preceded by Dan Quayle's often ridiculed concern for "family values") to frequent references in political interviews and talk shows. I have always felt that the safety movement should have been more cognizant of what was going on around it in order to contribute to accident prevention and to the betterment of the human condition. So I began to ponder what this talk of values might mean to loss prevention since there was very little explanation of what was meant by values in any of the above contexts, with the possible exception of the Eyre book.

I concluded it may be close to the foundation of how we should be managing safety but I wasn't sure how. In conversations with colleagues and clients and through further reading, I seemed to prepare myself for some inspiration on the subject. My casual conversations led to an invitation to speak on the subject of values-driven loss prevention at the June 1995 American Society of Safety Engineers Professional Development Conference (PDC) in Orlando. That led to a call from Alex Padro, the Acquisitions Editor for Government Institutes, Inc. Alex apparently had been interested in publishing a book on a values-driven approach to managing safety, health and environmental affairs. He asked me if I would be interested in writing such a book. At first I demurred, indicating that I had no book writing experience and probably didn't have the time to acquire it at this point in my life.

Then, I experienced a rush of stimulation that prompted me to ask if Alex was going to the PDC event that had precipitated his inquiry to me. He said he was. I suggested that he attend my presentation and, if he liked it and was still interested, we could discuss the details over lunch. He accepted, and I was relieved that I had not gone against what I often touted as my life script of being positive, keeping doors open rather than shut.

The stimulation, enlightenment, and adrenaline rush really started at that time. I found myself clipping articles that referenced values, reading related books, and taking notes on every life experience that had a kinship to this subject. All the while I talked to all my friends about my emerging thoughts. My wife was supportive, but during thirty-three years of marriage, she had seen several of these obsessive bouts and was just going along for the ride.

As I prepared for the PDC presentation, I realized that with some fleshing out and a little organization, I had enough material to give a two day seminar. I tried to condense and summarize during the presentation, but I'm afraid my talk ended up hurried and confused. Nonetheless, it was well-attended and well received, judging from a very positive evaluation and the number of business inquiries and warm discussions that ensued afterward. After the presentation, a lunch meeting with Alex set this work in motion.

So, I guess the short answer to my qualifications is that I have a publisher. The long answer cannot be my track record as an author. It must be that here is an idea whose time is at hand. But, let me add some qualifications to encourage you to read this work carefully.

I have been a safety professional for over thirty years. I passed the first-ever Certified Safety Professional exam. About twenty years ago, I wrote most of the Management Aspects section of the exam. I have done everything from inspecting factories for the Factory Mutual Engineering Division to managing a multi-dimensional loss prevention department for a complex Fortune 100 company with ten business units. In addition, I have served as president of The American Society of Safety Engineers (ASSE) and The American Society of Safety Research. Most importantly, during those assignments, I was fortunate to associate with some of the best and brightest people in this business and in American corporate leadership. I recognize many of them in my acknowledgments.

I have had articles published in *Professional Safety* and in other technical publications and edited *Readings in Safety Management*, published by ASSE. My writing has been infrequent and undistinguished.

These experiences have done little to prepare me to write a book on managing safety on a foundation of values. However, I've been a father to four animated and creative offspring and interact with three over-achieving and wholesome sons-in-law and the seven grandchildren they have produced to date. But serving a congregation of over four hundred wonderful people in the Glens Falls Ward of The Church of Jesus Christ of Latter-day Saints as their bishop was, perhaps, the most useful experience I have had. A few weeks into my five year service, one of my daughters asked, with the concern of someone who loved me and was aware of the pressures of such a responsibility, "How are you doing dad?" My answer was by reflex. I said, "Jean, during the last seven weeks I have learned more about what is important in life than I did in the first forty-eight years." It was a heartfelt answer that I could have defended then and could substantiate even better now.

As significant as my professional experiences have been, why would I focus on a relatively brief, part-time ecclesiastical adventure as a qualifier for authoring a book on managing safety driven by values? As an LDS bishop, one is exposed to the full range of domestic and spiritual afflictions affecting well-meaning people who readily confide in you. In brief, you get to vicariously experience the lives of hundreds of people and families in a relatively short period of time.

What did I learn from this? In short, I learned that what people believe, i.e., what their values are, determines their behaviors. Since behaviors have consequences, I found that if I knew what people believed and valued, I could predict their behaviors and, therefore, the outcomes of their pattern of living. James Allen explored these relationships in his *As a Man Thinketh* by using a seed/planting/harvesting allegory similar to that mentioned previously, comparing our beliefs and values to seeds and our behaviors to the fruits.

All this points to the fact that there are scientific and social laws that have been decreed by or rooted in a source that is recognized by most of us; this source is given a wide variety of names by different segments of the human family. For the purpose of this work, it is not important what we call the force but it is important to accept the concept that life has purpose and that almost everything happens by design and is not random. In the Prologue, I will postulate several "facts" that serve as the foundation for the material I will present. If that doesn't excite you, I'm almost certain the maturity grid in Chapter 18 will.

ACKNOWLEDGMENTS

Without the encouragement, tutoring, and editorial work of my son-in-law, Eric Poulsen, this book never would have come into being. In June, 2000, Eric completes his residency at Duke University Medical Center for a career in ophthalmology. For the past seven years, Eric has edited much of my work, and for part of that time has been a student editor for the *Journal of the American Medical Association* (*JAMA*). At times I think he knows more about what loss prevention is *really* like than I do.

The full support of my wife, four children, and two other sons-in-law has provided much needed stimulation for this daunting task.

The key figures in my professional development were Jack Snyder, who literally taught me what a safety professional was, and Ted Mullins, who gave me the opportunity to flower at Cheesebrough-Ponds and was always supportive, and, perhaps, the best manager I have ever encountered. I learned daily from Ted's example.

At Cheesebrough-Ponds, almost all the prominent people and some of the not-so-prominent patiently taught me and served as advocates. Bob Bennett, Dick Scheifele, Lloyd Giardino, Ken Lightcap, T. R. (Dick) Larsen, and Ralph Ward deserve special recognition.

The people who worked for me and with me—those who made me look good—should *all* be named. Space doesn't permit, so I'll just name one who exemplifies all their dedication and responsiveness. To this day, almost a decade after leaving Cheesebrough-Ponds, Lew Booker still frequently helps and encourages me.

During my service to the American Society of Safety Engineers, a host of fellow safety professionals inspired me and not only led me to the good I was able to accomplish, but rounded out my character and professional capacities. John Russell prepared me for the presidency and prepared the presidency for me. Chuck Culbertson worked tirelessly *for* me. I will never be half the safety professional he was. Tom Reilly prepared my Fellow application and was a relentless fellow traveler and example of all that is good in our profession and in the ASSE. Jim Smirles was always a source of wisdom and, by my reckoning, the nicest person on our planet. Fred Manuele sharpened my intellect by always asking good questions and providing unique insights. Charlie Dancer never declined to help me and has always been one step ahead of me. They exemplify all those who are too numerous to mention, those who only come after my family and church when it comes to warm feeling and reflections.

Those I have served under in my church callings and those I have served have done much to brighten my life and focus my priorities. Rod Hawes, Ken Theiss, K. B. Rasmussen, Hyde Merrill, And Dale Ensign have taught me so much that it scares me to think of what manner of man I would be without their influence. My current leader, R. Noel Hatch, New York Utica Mission President and former corporate trial attorney, teaches me new words and inspires new thoughts every time I hear him speak.

Of my clients, Bill Driscoll and Craig Bennett stick out as friends, benefactors, and people who made me feel good about what I was able to do for them.

My mom and dad and brothers, Bob and Wes, always encouraged me. No one could have had a more supportive family.

PROLOGUE

I view this book as a long open letter to safety professionals, business leaders, and others who preside over enterprises as vast as government agencies or as small as family units.

The world is experiencing troubled times.

Most of the answers suggested by leaders seem to deal with symptoms. Few people see the leaders as credible. Is there a way out and who can help us find it? I suggest that there is a way and that it may need to be exposed by some unrecognized and unlikely sources. Take a few minutes and look carefully and deeply into the mirror. I'm going to advance the premise that safety professionals have the answers but don't know it and are often ignored anyway.

As a world, our moral and ethical compass is badly out of adjustment, and until we calibrate it, there is little hope of finding our way out of our dilemmas. In order to calibrate, we need an accurate source. Most sources have been corrupted by greed and/or self-interest. Since safety professionals are not generally held in high esteem, we have not been a target of those who promote self-centered purposes. Hence, we have usually avoided the moral misalignment and ethical disorientation that have become epidemic. With a generally pure purpose, we could be the most qualified group to lead society in a new way of looking at reducing undesirable outcomes. On top of that, the skills and techniques we should use can be applied to bring to pass enriched sub-units, then larger groups, and eventually all of society.

The ultimate sub-unit is the family. It is the social building block of society, teaching the concepts that foretell success or failure of larger units, from schools and businesses to whole countries. The family also serves as the laboratory for experimenting with social change; it is the pilot plant for social methods and a training site for unit members, particularly the younger ones. And so, I will frequently use family analogies and examples in this book. Everyone is part of a family, sometimes several families, and can therefore relate well to such references and then extrapolate.

Most successful enterprises have mimicked time honored family tactics and often even refer to themselves as a family or extended family. We see it often in athletic teams and small business units and less frequently in larger groups. The leaders of such organizations take on the role of family patriarch. Each individual supervisor in a company has the potential to fulfill the father/mother image. If they do it well, they get the most from their subordinates. If not, ineffectiveness in one form or another results.

The culture of a team, organization, company, or whole society will predict their behavior and the results of their efforts. The culture will be determined by values taught by leaders/parents. The three best ways to teach are first, by example; second, by example; and third, by example. Organizational values must be developed; and skills must be learned and applied to transmit the values. Practices of leaders are central to success and a result of *their* values.

Almost eighteen years ago, my wife and I were invited to attend a party where all the attendees held common religious beliefs that we did not share. We were impressed with the wholesomeness of the discussions and the priorities expressed by the attendees. We discussed our feelings at length with each other and then dismissed the experience. Shortly thereafter, one of the people who had been at the affair asked if we had enjoyed ourselves. We said that we had and related our observations. That person was an attorney, Bob Alsop, who had employed our children as babysitters.

He asked if we would be interested in the religious beliefs of the group and I responded that I had my own and no thank you. I did express interest in what caused the group to behave in such a pleasing way, creating a warmth and good feeling that I had rarely experienced. He said that it was related to their religious beliefs, and I repeated that I was satisfied with my own religion and wasn't looking for a new set of beliefs. He gently suggested that my quest to find out what made those people "tick" would probably go unfulfilled. He had used the right bait and I investigated their beliefs. The correlation quickly became obvious.

This experience and many others has led me to feel that the loss prevention process will be most effective if driven by a set of values based on beliefs. The values can be developed by wise and selfless consideration of enterprise (extended family) character and needs. The enterprise will then be prompted to develop the skills required to extend the values throughout the organization.

Safety is a common ground and is non-threatening, so resistance should be minimal. Once the values and techniques are learned, applied, and installed in the safety arena, they can be extended and refined to enhance quality, productivity, or any other area of enterprise concern.

For-profit organizations would be the most hospitable environment to demonstrate the tenets I will discuss. In a capitalist system, they tend to be the most highly motivated elements of society. Once the concepts work there, government, education, and other organizations will see the merits and, hopefully, adopt the techniques and tactics that have made companies more profitable. Effective application of these concepts is designed to optimize profits, a primary driver which will always exist where the objective is financial gain.

An infrequently recognized fact is that better safety can improve profits ten percent or more and may be the single most important factor in business success. I will demonstrate that later and have done so in greater detail in professional journal articles for human resource professionals and risk managers. In his presentation of the "five pillars" on which an economic policy must rest, Peter Drucker includes conservation of human and man-made assets. This visionary management guru, who may never have been in the presence of a safety professional, recognizes the vital role they can play. Some company CEO's are aware of it too, but they are unfortunately few in number.

"Value Inspired Resource Optimization", or VIRO, is the name I have chosen for the process this book will expose, discuss, and detail. What we are talking about is making the best use of the physical and human resources of an organization by reducing and eventually eliminating undesired outcomes that deplete hard earned assets. This is accomplished by establishing a base of beliefs and values that predict success.

When I use the term "loss prevention," which I like best to describe what safety professionals should be doing, I mean the following: *the conservation of our physical and human resources by protecting people, property, and the environment from undesired outcomes that depreciate them or inhibit their enrichment.* Although this terminology has been tainted by people and organizations who have minimized its meaning rather than magnified it, I think loss prevention best describes what I am talking about in this book. So I will continue to use it with the definition suggested above. Please focus on the ideas rather than the terminology.

The question of why the worth of safety professionals has gone largely unnoticed and under-appreciated needs to be answered so we can apply the theories and techniques detailed in this book. If we can understand the importance of directing any outcome enhancement efforts through a foundation of values stemming from beliefs, we can apply the techniques to our loss prevention efforts with the enthusiasm that is required to effect change.

WHAT IS THE LARGER VISION?

All undesired outcomes are a form of accident. The techniques that are employed to minimize accidents and injuries can be used just as effectively to enrich our society by optimizing our physical and human resources. Finding the "root causes" of failed systems that rob our resources is just as essential in improving health care, education, civil order, and race relations as it is in preventing a single accident. The goal is to develop processes predisposed to success instead of failure. Starting with a foundation of values based on beliefs that serve to direct efforts will work in both spheres. Loss prevention is an ideal venue to demonstrate the concept and develop models. These can then be applied to everything from product quality improvement to parenting.

WHY HAS THE POTENTIAL OF SAFETY PROFESSIONALS GONE LARGELY UNTAPPED?

There are numerous reasons, but I think the three most crucial ones are as follows:

1. Safety professionals have been poor at public relations and self-promotion. They are certainly not their own best advocates. Every profession, organization, and group that has gained recognition has employed strenuous efforts to promote themselves, including doctors, lawyers, accountants and even professional athletes, not to mention tobacco and alcoholic beverage industries. In the late 1970s and early 1980s, as an officer and later president of the American Society of Safety Engineers (ASSE), I championed the creation of a public relations function at society headquarters. Some enlightened leaders supported this effort, but the proposal met stiff and determined resistance. The cry, "We are not in the public relations business," was heard often. The truth is that everybody is in the public relations business whether they choose to recognize it or not. The ASSE has done some promotion of the profession but has only scratched the surface of what can and should be done.

2. Self-serving, well-placed, contrary interests construct road blocks that have rarely been breached. Loss prevention has historically been driven by the insurance industry and more recently by regulatory interests. In his book *The Invisible Bankers*, Andrew Tobias describes what really motivates the insurance industry. The euphemistic title reveals that it is not loss prevention.

If losses can be predicted and controlled, the insurance industry and all its parts, from the companies themselves to brokers and risk managers, lose. So, they have little interest in serious, determined efforts to devise a system that will predict cost minimizing outcomes. The advocates for that mindset are at times even seen as pariahs in the insurance and industrial communities and excluded from involvement in critical meetings and activities.

Regulatory interests focus on lawyers and regulations that mask the real problems. They often subvert serious prevention efforts by overwhelming them with largely irrelevant minutia. They take on a life of their own and are highly resistant to corrective measures.

3. Executive ignorance (for lack of a better term) has been the third strike against safety professionals. Business schools teach no safety and not much insurance. Executives tend to trust their instincts when they encounter subjects in which they are not well-schooled. Personal insurance, such as car, home, and life insurance, may be referenced but will cause even an intelligent executive to reach wrong conclusions. Those costs are largely uncontrollable and astute, competitive purchasing is usually the best way to minimize them. Just the opposite is true of the highest cost industrial insurance, which *is* loss sensitive. The best way to keep costs down is to have an enviable loss history.

These business leaders also have a bias for traditional approaches. Brokers and risk managers are steeped in tradition, were often educated in the same places as senior managers, and often belong to the same country clubs. Breaking this cycle of dependence has not been accomplished often. Self-insured entities probably represent the most enlightened industries when it comes to a healthy perspective on funding and preventing losses. The current need to do more with less and the inclination to abandon costly tradition to remain competitive may help this timely idea avoid a premature dismissal.

HOW DO WE MOVE FORWARD?

Given this setting and background, how can developing a twenty-first century safety culture, enhancing profits, making our businesses more competitive, and teaching concepts that will enrich enterprises as well as help us re-invent government, solve the nation's health care problems, overcome our education deficiencies, and improve family life (which is the heart and soul of our society)?

This proposal demonstrates that outcomes emanate as a natural consequence of the beliefs and values of a social setting that incubate and produce the outcomes. The term "corporate culture" is how most current business books refer to this complex phenomena. The critical conditions and behaviors that foretell unwanted outcomes can be viewed as "indicators," and can be measured and managed while preparation begins to get at the real root causes or "predictors," (which are the values held by the population) as they relate to preventing losses. The value base of the organization can then be enriched and progress and results measured. The knowledge of the root causes can then be fine-tuned, and methods applied to other areas of concern with the population understanding the important correlations and the validity of the methods.

"Will this work?" my wife inquired as I excitedly shared my thoughts with her on a boat ride on Kattskill Bay in Lake George. "Of course it will," I confidently responded. Then she asked the question that I realized I would have to answer before most people would take this seriously. "Where has it been applied and proven to be effective?" The answer to that is both everywhere and nowhere. It is being proven every day but is not recognized or understood. I will spend much of this book trying to convince readers of my thesis.

THE PREMISES

The validity and power of these ideas depends on the fundamental assumptions listed below:

1. We are governed by natural and social laws that transcend human laws. The social laws may be more difficult to quantify but are no less irrevocable or immutable. Accepting that absolute truth exists is imperative. People tend to understand and accept scientific laws but have more trouble with uniform acceptance of social laws, even if they prove themselves with the same consistency. Perhaps it is because the cause and effect are not as timely and precise. Also, the violation of some social laws can produce instant gratification that yields to long term unhappiness which is so far away that some people do not concern themselves with it. In a corporate setting, embezzlement, using sophisticated electronic techniques, is a classic example of ignoring social laws while applying natural laws. The result is never good.

This is both a problem and an opportunity. It is an opportunity because it has not been adequately explored, hence this book and this discussion. Think about instances where dishonesty was practiced. Did that ever yield a positive result or outcome? How about infidelity? Could the long term result of a partner cheating on their companion have a positive outcome? Could citizens vote to repeal the law of gravity? Of course not, and neither can citizens repeal social laws. So they must learn to live with them and use them for their benefit and happiness.

2. Beliefs and values predict outcomes. This is the basis for this book and the *keystone* premise. Years of experience in the different dimensions of my life have given me a fervent belief that this is true. If I know what people believe and what they place value on, I have found I can predict behaviors and, in turn, outcomes. If you reject this out of hand, this book will be hard reading for you. If you accept it, I think you'll find the book stimulating and enriching. If you're not sure, I think there is a good chance that I can convince you in the course of unfolding my story.

If you believe in chastity, are your chances of being victimized by a sexually transmitted disease increased or decreased? If you believe that you are entitled to anything you can get your hands on as long as no one catches you, are you more or less likely to wind up in jail than someone who believes stealing in any form is wrong? These may seem like "no brainers" but a lot of people don't seem to have gotten the point. Fewer yet have applied the point the way it will be suggested in this book.

3. Beliefs can be influenced and values taught. If this is not true, we are all in a lot of trouble. We are caught in a world that is beyond our comprehension, is rudderless, and defies reason. Beliefs *can* be influenced; values *can* be taught. Consequently, conversions to religious, scientific, and political belief systems take place every day. Democrats become Republicans and Catholics become Muslims and the reverse.

So values can certainly be taught and learned, which is why we see dramatic changes in what a society values within a generation. The big question is not whether we can, but how it is done and with what degree of precision and over what time frame. I may not provide a precise answer in this book, but I intend to provide some material for contemplation and some techniques and methods.

4. It is better (and easier) to shape outcomes far upstream than to react at the last possible moment. Anyone who has tried to control an unruly group of people (ranging from a civil disturbance to a lot of people talking in a room) knows that they would have preferred to have dealt with the first talker or the first person who stepped out of line.

When the Russians moved to an open society, I doubt they anticipated the emergence of a Russian Mafia. They not only have freedom of choice but some anarchy to go along with it. If they had anticipated what is happening, it would have been easier to institute measures to discourage crime than to react now, when things are out of control.

If you work downstream, you deal with symptoms, and solutions often only mask the real issues. Consequently, the real problem goes unchecked and often deteriorates. Shaping the flow of the trickle upstream is always easier than trying to influence the torrent downstream.

5. Solutions to problems depend on accurate diagnosis. Medicine offers a helpful illustration. An elderly gentleman visits his doctor after becoming severely short of breath. Many different medical conditions could be responsible for this symptom, and a thorough physical exam, appropriate blood tests, and possibly x-rays, may be required for accurate diagnosis. Appropriate treatment hinges on diagnostic accuracy. If the physician's diagnosis is wrong, the therapy he or she recommends may worsen the patient's condition or, at best, simply not help.

6. The process used to minimize accidents/loss can be applied to optimize desired outcomes in areas ranging from product quality to individual lives. The side benefits of this approach could range from the icing on the cake to winning the lottery. I believe it has the potential to be a lot closer to a big pay-off that is not only enabling for loss prevention interests, but enchanting for global thinkers.

When I first read Crosby's book *Quality is Free*, I concluded that he could just replace the word "quality" wherever it appeared with "safety" and reissue the book with a new title, *Safety is Free*, and have another best seller. That was my first significant exposure to the quality revolution almost ten years ago. Since that time, I have taught a short course for safety professionals on the parallels between the safety and quality processes and what they can mean to enhanced loss prevention.

The bridge I have identified between the disciplines is not technology, although there are some similarities. Nor is it management methods, although those offer even more parallels. Rather, it is values and how they can be used to support both processes and produce synergistic results.

What this means to me is that if the theories and processes exposed in this book work, they can be applied to quality and in turn to virtually every area in which we are concerned with results and desire to avoid outcomes that are displeasing to us.

GOALS

This prologue would be incomplete without detailing what we intend to do together in this book. You will need to participate in order to maximize your benefit.

1. Substantiate the premises presented in the text.
2. Suggest and explain values and beliefs that predispose an environment free of undesired outcomes.
3. Not only explain the beliefs and values but indicate why they are important to minimize loss.
4. Provide a reasonable system to measure organizational values.
5. Detail how a group can move toward values that will limit loss and enrich the organization.
6. Provide ideas and stimulation techniques that will allow the value focus/enhancement to be correlated with outcomes/losses.
7. Offer convincing evidence that this can and will work if applied correctly.

CAVEAT

Before we go any further, there is one more facet of this introduction that must be made clear. I have strong religious beliefs and they are firmly rooted in a particular faith. Support of my theories and a desire to act on them in no way requires readers to share my particular set of beliefs. But, they must hold to some basic tenets that are common throughout the world. Hopefully, readers will emit "ah hah's" as these ideas are examined. They will rekindle deep feelings that are part of our humanity, but which we ignore much of the time.

Our creator has given all of us a sense of right and wrong and instincts for self-development. Part of our self-development instinct is a strong desire for self-preservation. People with different cultural/religious roots call it different things, but all of us know it is part of our make-up.

I am convinced that many bestselling books such as *The Seven Laws of Spiritual Success* by Deepak Chopra, *Embraced by the Light* by Betty J. Eadie, and *The Fourth Instinct* by Arianna Huffington appeal to our most latent characteristics. I will borrow from them and similar works to demonstrate the prescience of my theories and how we don't need tangible evidence to know they will work.

Although I hold tenaciously to my particular set of beliefs, I am absolutely convinced that our creator(s) could have forced a single religion upon us and did not do so for reasons that will become apparent at some later time. My particular religion allows all men and women the privilege of worshipping according to the dictates of individual conscience—how, where and what they may. I am very comfortable with that concept and find the variety of beliefs to be very interesting and pleasing, like looking at a kaleidoscope's many colors rather than a single color. I do find it fascinating that virtually all major religions hold to similar dogma relative to how we should treat each other, how we should organize socially, how we should regard deity, and the transcending potential that we all have. My values-driven approach to loss prevention aims to capitalize on these common, overarching principles.

SECTION I

INTRODUCTION TO VIRO

1

RELEVANT HISTORY OF THE SAFETY MOVEMENT

*"He that is too secure
is not safe"*

The safety movement has little relevant history; it primarily offers an opportunity to learn from our mistakes. That will be the theme for this chapter and will recur throughout the book.

Careers in safety really didn't exist until the twentieth century. Well, that's not completely correct. Safety is intrinsic to nature, as evident in animals' camouflage. But the first human safety engineers may have been the cliff dwellers who pulled up their ladders into the cave each night or biblical homeowners who were cautioned to build parapets on their roofs to prevent falls. And the parents of the world have always been in the safety business. And by right they should be.

Injury in America/A Continuing Public Health Problem, (National Academy Press, 1985) produced by the Committee on

Trauma Research, Commission on Life Sciences, National Research Council, and the Institute of Medicine, reported, "Injury greatly surpasses all major disease groups as a cause of prematurely lost years of life, because it is the preeminent cause of death among children and young adults." Note that this group studying injury in America included *no* safety professionals. Most injuries happen in the home. The homemaker is the first line of defense. Home exposures range from exotic cleaning agents to high hazard "toys" and drug abusing teens, so parents must be ever vigilant to protect their children and spouses. Without my wife, who is the best safety practitioner in my home, I hate to think of what may have become of me.

It should be noted that a May 19, 1993 supplement to *USA Today* reported on "Safe Kids in America" and indicated that injury is the "Number One Killer of Kids." AIDS is fast becoming the leading cause of death in the twenties to thirties and may have surpassed trauma for this age group or is at least getting close. Of course, AIDS also fits my definition of an accident.

While parents have become well aware of the growing importance of safety, industrial leaders have been remarkably resistant to learning on this subject. Part of that fault must lie at the feet of safety professionals. They have tried, with publications such as *Scope and Functions of the Profession* published by the American Society of Safety Engineers (ASSE), but to date the world little recognizes the potentially beneficial role safety professionals can play.

The Board of Certified Safety Professionals (BCSP) has broadly defined the profession with comprehensive exams that cover an ever-widening group of specialties. Graduate and undergraduate programs have become accredited at major educational institutions with capable faculty. Those programs offer a wide range of courses that broadly define the profession, and they grow yearly. *Professional Safety*, the journal of the ASSE, publishes articles covering a continually broadening spectrum of subjects. The society has adopted a motto that more comprehensively defines the profession. It indicates that the profession is involved in the protection of people, property, and the environment, based on a statement first suggested by Bill Nebraska at a long range planning meeting in the early 1980s.

With all the work of ASSE, BCSP, and their publications, one would think that the world, particularly the industrial world, would be utilizing all the potential of the profession and harnessing it to optimize their profits. Nothing could be further from the truth. For the most part, the concept of the safety practitioner as someone who makes inspections and oversees the distribution of safety glasses and safety shoes lives on.

The BCSP was established over twenty-five years ago. It led to the certification of safety professionals, and that prompted licensing in many states. Not long after, the ASSE began certifying and accrediting safety curricula. The Occupational Safety and Health Act of 1970 brought safety into the regulated realm. Note that the better part of a couple of decades passed before a safety professional headed OSHA: there *is* a message here. Ironically, in some ways, OSHA depreciated safety by its narrow and somewhat pedantic definition of the field. But it did get the attention of industry leaders.

Other significant history of the profession includes the first workers' compensation legislation (prior to World War I). That dissolved common law defenses for work injuries–a mixed blessing. Insurance companies became the leaders in safety management–another mixed blessing. Then in 1936, the Walsh-Healey Public Contracts Act was passed. It was interpreted to mean that government defense contracts required some evidence of attention to accident prevention—yet a third mixed blessing.

Moving back to post-OSHA, we see the Consumer Products Safety Act, focusing on the likes of Christmas tree lights, matchbook covers, bicycles, and hair dryers. Together, the CPS Commission, the National Institutes of Occupational Safety and Health, and the Food and Drug Administration have declared many of life's pleasures hazardous to our health. There is little or no cost-benefit analysis, and safety professional involvement has been minimal. Where does all this leave the public? Confused and ready for attitude and perspective adjustment. When and how and who will do that? Safety professionals should do it now. The concepts in this book can lead the way to a comprehensive definition of the safety profession and bring light to a dark and cloudy landscape that has been sullied by legislation (lawyers), well-meaning but misdirected do-gooders, self-centered and insulated business leaders, and an insurance industry that is justifiably focused on investments and profits rather than safety.

Okay, what mistakes have been made by safety practitioners and how can we learn from them and correct them?

- **We've allowed others to define our profession.** Safety professionals need to find and use forums to tell their story. When a safety issue makes it onto the national scene, safety professionals should be on television and in the media being consulted.

- **We've been satisfied with a subordinate role.** The name of the game is prevention. That is what safety professionals do. They should be front and center in conserving physical and human resources. Others should not crowd them out. Who introduced behavior-based safety? Not safety professionals. Why not?

- **We spend too much time talking to each other, rather than the rest of the world.** Safety professionals need to widen their horizons and, through all the media, speak to the world. They need to listen more carefully to what is happening in the world and relate it to how they can contribute; then do it.

- **We don't think much of ourselves.** Safety professionals need to enhance their self-images and define and place people in leadership positions who will speak for them and find listeners.

- **We have never developed advocates.** Safety professionals need to provide input to influential people and encourage them to speak for the safety industry and for what it does. They need to make friends with people like Sam Donaldson, Stephen Covey, and Newt Gingrich.

Unfortunately, safety practitioners have already missed innumerable opportunities to rise and shine. The good news is that opportunities are greater than ever before. Safety professionals are stronger and smarter than they have ever been (they're better prepared). VIRO is a vehicle that can be used to clearly expose the things that they can do. The full range of benefits to employers is revealed by a values-driven approach to loss prevention.

2

THE CURRENT CLIMATE OF RELEVANCY AND RECEPTIVITY

"We may affect climate but weather is thrust upon us."

When preparing for a trip, it is a good idea to check the weather. The VIRO trip would call for a certain conducive climate. Climates are predictable; but weather is often unpredictable and variable. Readers shouldn't concern themselves with the weather or isolated reactions to this concept. What *is* important is the climate; what is going on among the "movers and shakers" of the world.

This chapter describes the climate in which we live and what conditions are needed for VIRO to "sail." Most people wouldn't try to sail without wind, nor go sailing in a hurricane. By the end of this chapter, it should be obvious that the current business, political, and social climate is ideal.

This chapter will first list and explain the necessary "environmental" characteristics. Then it will comment on what is happening in the business and political arenas. It will look at business and political writers; briefly restate their themes; then reflect on the impact they seem to be having. To probe the subtleties of the social/political/business environment, this chapter will review recent popular non-fiction works that have been widely accepted, but more in the social rather than business or political realm. Lastly, it will look at the relevancy of all this thinking to loss prevention in general and specifically to the concept of managing safety on the basis of a foundation of values.

Acceptance and implementation of a values-directed approach to safety requires the environmental characteristics listed and explained below.

Environmental Characteristics Required

1. A need must exist.
2. Capacity to understand is fundamental.
3. Explanations must be clear and palatable.
4. Cost must be acceptable.
5. Perceived benefits must exceed the cost.
6. Culture compatibility is essential.
7. Drivers need sufficient power and intensity.
8. New ideas need champions.
9. Complete preparation is a must.

ENVIRONMENTAL CHARACTERISTICS

1. A need must exist. Most people have seen in action the proverb, "Necessity is the mother of invention." Further, few products or inventions can sustain themselves unless there is an understood need. If the need exists but is not generally recognized, nothing will happen. If someone is going to build a product or expose an idea, they should be able to see the need and know that others see it too or be confident that they can display it to others in a way they will understand. Who would have thought they needed the light bulb or the fax machine?

2. Capacity to understand is fundamental. It is difficult to explain something to someone who doesn't have the capacity to understand. Jean Piaget, the Swiss psychologist who did pioneering work in child development and learning theory, discovered that even many adults don't mature into abled abstract thinking. The stage of comprehending ability must be appropriate.

3. Explanations must be clear and palatable. This places a burden on the messenger. At times, there are a lot of messengers trying to tell essentially the same story. Usually, that tends to create confusion. Each one sees the message a little differently, and hence tells it from their own perspective. It's a little like all the Christian churches who each have their own version of the same message. They all try to edify, but, cumulatively, confusion often develops.

4. Cost must be acceptable. Many people would like to go to the moon. Only a few have had the experience. If new ideas are going to gain wide acceptance and practice, they must have a cost that is within the reach of more than just a small segment of the population, no matter how useful or exciting the idea.

5. Perceived benefits must exceed the cost. Big companies have sophisticated ways of evaluating investments. They stipulate pay back periods based on interest rates and their resources; new investments must have a return-on-investment that meets their criteria. Individuals do the same thing, albeit with criteria that are usually less sophisticated and often have an emotional ingredient.

6. Culture compatibility is essential. The seed sown on stone won't grow. That is no reflection on the seed. Try introducing a food or beverage into a culture that finds it to be displeasing. Worse yet, try a new idea that contradicts a set of beliefs that are staunchly held by the group you are attempting to influence. It would be difficult to market electric appliances in an Amish community.

7. Drivers need sufficient power and intensity. Sometimes good ideas just can't gather sufficient financial or moral support. Without sufficient support, no matter how good the idea, it will sometimes "die on the vine," only to be reborn someplace else when it finds enough interest to sustain it. Good ideas often get started only to die for lack of commitment or action.

8. New ideas need champions. No matter how good an idea is, if the wrong person is an advocate, it is stigmatized by the person's reputation. It's the old messenger problem. Recently, Louis Farrakhan, leader of the Nation of Islam, led a march on Washington to empower black men. Very few people took issue with the basic message. Many leaders, even in African-American communities, took substantial issue with the messenger. The impact was not nearly as great as it could have been with a totally credible messenger.

9. Complete preparation is a must. If you sow seeds before the field is plowed, when it is full of weeds, you may grow something, but differentiating and untangling the good plants from the weeds will make harvesting very difficult. Someone or something must prepare the way. The new idea needs to emerge at the right time, in the wake of a preparatory message.

These environmental characteristics have a negative air about them. But when the chapter is completed, the reader can catch a vision of how our climate may have adjusted to perfectly accommodate the growth of VIRO.

WHAT BUSINESS, POLITICAL, SOCIAL, AND SPIRITUAL WRITERS ARE SAYING TO US

For years, Peter Drucker has been a prolific writer. He is often viewed as "The Man Who Invented the Corporate Society" and the father of American management. More recently, Tom Peters has become widely read and now speaks to large groups of managers who are trying to achieve a competitive edge. As a category, writers and consultants on quality control have enjoyed a growing audience. Edwards Deming and Philip Crosby are two of their most prominent spokespersons. Michael Hammer triggered a firestorm of reengineering with the book he wrote with James Champy. Change can't be successful without considering the environment. These men and their writings are shaping the modern business environment.

Politics has a profound effect on business as well as the social fabric of our country, and these three areas are becoming progressively more intertwined. Witness the current criticism of large media companies over the kinds of music they sell and the nature of the movies they make. Afternoon talk shows often end up being referenced in political talks, even messages from President Clinton. This probably isn't an aberration. The public will be seeing more politics, not less, in the future. The section of this book on political writers references *To Renew America*, House Speaker Newt Gingrich's recent book; *Values Matter Most* by noted author and avowed liberal Ben J. Wattenberg; and *The Agenda*, Bob Woodward's exposé on the Clinton Administration.

William Bennett, who was a drug "czar" under George Bush as well as the Secretary of Education, seems to show up on almost every panel on political/social issues. He's bridging the political-social gap. He is a regular on C-SPAN and has written a book on virtues. On *Good Morning America*, he said he used the word virtues instead of values in his book simply because values was overused. They don't show up as synonyms in my thesaurus. Do they in yours?

For a "climate check" on social commentary, we'll turn to Deepak Chopra, who is widely read, and can be seen on cable television. His recent book *The Seven Spiritual Laws of Success*, is a best seller. I also review work by Richard Eyre—a business consultant and author who has enjoyed great success writing about parenting and social mores—and Stephen Covey—a former college professor and currently a bestselling author of self-help and business books and head of a very successful consulting firm.

These people as well as many others are shaping the attitudes of the most intelligent and influential people in our society. What they are saying is important and relevant to how safety can be managed and specifically to the new concept of managing loss prevention with a foundation of values. What follows are some personal insights and commentary on what they are saying and how their ideas relate to the subject of this book and form the essential environmental characteristics that can predict the future success of VIRO.

Business Writers

Tom Peters started his rise to fame with the landmark book *In Search of Excellence*, which he co-authored with Robert Waterman. He followed with a sequel and then moved on to titles like *Liberation Management* and *Thriving on Chaos*. Rather than focusing, his later works are quite varied and probably too complex to be seriously followed as someone might follow a blueprint, although they are certainly thought provoking. At times in his lectures, he even counsels the attendees not to bring some of his ideas to their bosses, lest they jeopardize their jobs. Let's concentrate on his maiden work, published in 1982.

The book did benchmarking before many people called it by that name. In his introduction, he offers encouragement, suggesting that the best U.S. companies can compete with the Japanese. He recognizes that the best companies treat their people decently; ask them to shine; then cut out the formality and fat rule books, replacing them with first names, project-based flexibility, and everyone contributing. On page ten, he exhibits the McKinsey 7-S Framework (see page 16). The seven S's are structure, strategy, systems, skills, style, staff, and (you guessed it) shared *values*. All the S's are connected to shared values; none of the others are connected to more than five other S's. Shared values bind the attributes together.

Peters' work results in the emergence of eight attributes he feels are common to all excellent and innovative companies. They are: (1) A bias for action; (2) Close to the customer; (3) Autonomy and entrepreneurship; (4) Productivity through people; (5) Hands-on *values-driven*; (6) Stick to the knitting; (7) Simple form, lean staff; and (8) Simultaneous loose-tight properties.

In addressing conventional wisdom and the "conventional model," Peters says, "The rational model causes us to denigrate

McKinsey 7-S Framework ©

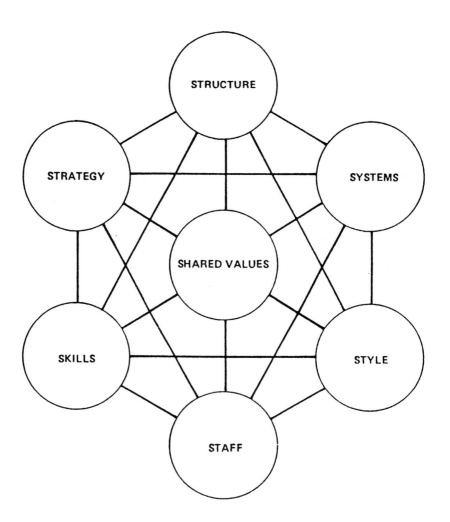

the importance of values." He follows that up by suggesting that in the excellent companies, "major decisions are shaped more by their values than by their dexterity with numbers." The book introduces a lot of new jargon, and much of it, like Management By Walking Around (MBWA), has stuck to this day. Much of the emphasis on culture during the last decade emanates from the stories in this book. Speaking of stories, Peters notices that in excellent companies, part of their culture is based on storytelling. More on that important element later. A typical comment reads, "Poorer-performing companies often have strong cultures, too, but dysfunctional ones." The heading "The Importance of Culture" precedes three pages on the subject.

Much of the book is devoted to stories about excellent companies and elaboration on the eight attributes mentioned before. Under "Hands-On, Values-Driven," Peters says about the excellent companies: "We are struck by the explicit attention they pay to values." In the first paragraph in the section, he says that if he was asked to distill one truth, he would reply, "Figure out your value system. Decide what your company stands for." Peters spends a page quoting Tom Watson, the father of IBM, who emphasized that success is tied to the beliefs of the organization—the appeal they have for the people–then faithful adherence to the beliefs. Later in the chapter, Peters lists the beliefs of the excellent companies. He says they are:

1. A belief in being the "best".
2. A belief in the importance of the details of execution, the nuts and bolts of doing the job well.
3. A belief in the importance of people as individuals.
4. A belief in superior quality and service.
5. A belief that most members of the organization should be innovators, and its corollary, the willingness to support failure.

6. A belief in the importance of informality to enhance communication.
7. Explicit belief in and recognition of the importance of economic growth and profits.

Now, what is missing from all this and what does it mean to safety professionals as they start to think about managing safety from a foundation of values? These are some observations:

- In spite of saying the *one* important *truth* relates to values, "Hands-on, *Values*-Driven" is the second shortest of the sections on attributes.
- The authors speak about others having trouble focusing on values because they are "soft"; then they fail to grab onto the reader.
- They talk a lot about culture and its importance, but don't relate values and beliefs to culture, nor say much about how to engender or measure any of them.
- Values and beliefs are discussed a lot; the use of the terms in management lingo and subsequent books and articles is increased; but, there is little or no hard information.
- Beliefs are listed generically, in contrast to the emphasis on individual cultures. They should have customized beliefs and values.

Deming and Crosby's work is also important to understanding the current ideological climate, but they are referenced extensively later when quality and safety parallels are discussed, so they are not discussed here. I once was puzzled by how quality and safety related to each other. I graduated to thinking of them as cousins; then brothers; later twins; finally Siamese twins.

Peter Drucker has written a virtual library of books. But let's focus on a fairly recent work, *The New Realities*. Here are some of the key points he makes, with an interpretation of how they apply to managing safety in a soon to be twenty-first century environment. Then let's relate his thinking to the values-driven approach to safety.

1. The foundations of success are education and discipline, not capital investment. Drucker is emphasizing the growing importance of human resources. A values-driven approach to loss prevention focuses on the people element.

2. Tunnel vision is the degenerative disease of specialists. The safety movement must operate off a broad base. If it doesn't, it will obsolete itself, in spite of its potential. The first rule of pathology states: *"Most well-trodden paths lead nowhere."* VIRO utilizes the broadest of bases and relates to every other aspect of organizational life...and, it inspires creative thinking.

3. If areas become politicized, service will degenerate. This speaks to turf battles and self interest instead of selflessness. VIRO encourages compatibility and unity of purpose. It builds bridges instead of walls.

4. Charisma without a program is ineffectual. The smoke and mirrors approach can work if the competition is not too sharp. That isn't the case very often today. The traditional methods of doing safety are outmoded. The new model should be values-driven.

5. Work will be done by task focused teams and organizations will become flat. This is happening more every day. Safety can't be managed by hierarchy. Remember Dow's Law: *"In a hierarchical organization, the higher the level, the greater the* *confusion.*" Safety responsibility must be assumed at the lowest level. VIRO engenders that better than any other approach.

6. Information is data endowed with intelligence. Most safety statistics are data and not information. If a capable manager can't instantly be led to an action plan by safety statistics, they're just data...and that is lamentably just what they almost always are. Drucker emphasizes the need management has for information, not just data. VIRO inherently provides real information.

7. Education can't be confined to the schools; it must be done and done well by every employing organization. What could be more important than teaching values, particularly when they aren't being taught in many homes and not at all in schools? VIRO addresses how to do it, how to measure progress and what results are expected.

Of all the recent management fads, the reengineering craze set off by Hammer and Champy reigns. *Reengineering the Corporation* is not an intellectually challenging book–it doesn't tell you *exactly* how to do what the title says; it isn't easy or fun reading and at times could even be called tedious. But it started a revolution of sorts. It clearly touched a nerve and addressed a concern. Management recognized that minor adjustments and mid-course corrections alone wouldn't guarantee their survival, much less their prosperity. What it is all about, by the authors' definition, is no more than "starting over." The process is no more than applied common sense. VIRO does just what reengineering advocates: it starts over with a new foundation.

Political Writers

The two biggest players on the political scene now are the president, Bill Clinton, and the Speaker of the House of Representatives, Newt Gingrich. In *To Renew America*, "Newt," as everyone has come to know him, tells what he believes. His thoughts are significant and relevant to changing the way safety is managed. Mr. Gingrich sets the stage by giving his version of the six challenges facing America. By way of introduction, he points out that there is no middle ground; society will either pull itself together or continue to decay. He says: "An America that arouses itself to replace the culture of poverty and violence and insists that its children learn the core *values* [italics added] of American civilization is an America that will find each challenge more invigorating than the last. But an America that remains passive and apathetic, divided and confused, will be on the road to decline."

Let's take a look at the challenges and what he says that is relevant to safety and, particularly, to managing loss prevention on the basis of a foundation of values.

1. "We must reassert and renew American civilization." The Speaker suggests that our civilization has spiritual and moral dimensions that form its foundation. He feels that since about 1965, a cultural revolution of sorts, led by "cultural elites," has tried to discredit our heritage and has effectively championed irresponsibility that would be anathema to our forebearers. He says we must reestablish a "legitimate moral-cultural standard," or a system based on our beliefs and values. A few specific quotes that set the tone are:

"America is an idea, the most idea-based civilization in history. To be an American is to embrace a set of values and living habits that have flourished on this continent for nearly four hundred years."

"In the mid-1960s, this long-held consensus began to flounder. The counterculture began to repudiate middle-class values."

Referring to the importance of stories about which Peters talks and this book will discuss in more detail later he says:

"America is a series of romantic *folktales* that just happen to be true."

He also references Dr. Deming, and by inference, the quality movement when he says:

"Deming can be understood only within the values that are at the core of the American experiment."

He pays homage to American spiritual/religious under-pinnings by quoting the eloquent and frequent references to deity by Franklin, Roosevelt, Jefferson, and Lincoln. Individual responsibility, free agency, or self sufficiency are championed.

2. "We must accelerate America's entry into the Third Wave Information Age." He emphasizes, as Drucker has, that schools should not have a monopoly on learning and that we must be fast on our feet and with our heads. America must renew the high value we have traditionally placed on learning.

3. "We must rethink our competition in the world market." Newt issues a call to action and a caution: "The best of our competitors are very, very good." He suggests that we need to get back to the value our society has traditionally placed on hard work.

4. "We must replace the welfare state with an opportunity society." He says we must abandon a failed welfare program and replace it with a new society, one that encourages self-reliance, not dependence. He suggests the poor have lost their faith in government and "want their children reeducated in core American values." The Speaker correlates drugs and violent crime with our welfare state and suggests that, "Establishing safety is the first foundation of creating opportunity for the poor."

5. "We must replace our centralized, micro-managed, Washington-based bureaucracy with a dramatically decentralized system more appropriate to a continent-wide country." Volunteerism and moving the decision-making and resources closest to the action, to grassroots values, are constant themes. He sees this as related to the assessment from the political scientist, de Tocqueville, who he is fond of quoting, suggesting that America is great because America is good and that when it stops being good, it will stop being great.

6. "We must be honest about the cost of government programs and balance the federal budget." Here the theme is prudence, responsibility, and concern for our children and grandchildren. Basic values are the focus of his explanations.

When he talks about The Contract with America he suggests that: "No reform effort of recent years fit the *values* of the country more than this bill." Not everyone agrees with that assessment. Time will tell.

As an educator, Newt speaks about Learning Versus Education in one of his chapters and suggests that "We are trying to woo people away from one set of cultural *values* (physical force, dependency, irresponsible childbearing) and into the more powerful and prosperous values of hard work and education." Then he gives one of the rare solutions found in all these discussions of *values*. He suggests that incentives give a lot of leverage and that government has been using them in the wrong way.

This chapter has spent a lot of time on Gingrich's book because he speaks so directly to values, has been so popular (*Time* Man of the Year) and represents the thinking of a large and seemingly growing segment of our population. Even as a former conservative Republican, I disagree with some of his ideas, as I see them as being incomplete and, in spite of all the talk of values, disconnected. Some of the appeals to religion attract religious zealots whose primary trademark seems to be intolerance. Some of the approaches seem to punish the innocent. America has always been a compassionate country. Compassion isn't mentioned often in some of the new agendas. While speaking of self-reliance and decentralization, some programs seem to restrict individual freedoms (tort reform) and free up groups to possibly plunder (regulatory reform and elimination). Some of the education reform ideas like vouchers seem good on the surface but could produce a two tiered system insensitive to those with the greatest needs.

The final analysis is that the new congress and its Speaker are talking a lot about values but at times seem as confused about them as William Bennett. There are times when demagoguery and name calling cloud the message. If we focused on the values and went from there, could this be avoided? Possibly. Maybe we could get there with VIRO.

Bob Woodward's book *The Agenda: Inside the Clinton White House*, suggests confusion and blurred values determining executive branch positions. Woodward is a credible political reporter. Remember *All the President's Men*? He, Carl Bernstein, and the *Washington Post* almost single-handedly caused the resignation of Richard Nixon. Of course, the root cause was Nixon himself. *The Agenda* results from numerous interviews and is filled with quotes and insider insights. For example, on the new chief of staff, Leon Panetta, as he was about to take a position: "Campaign aides had told him that Clinton was deadly slow to make decisions." Direct quotes from named people are used to support such conclusions. The theme of this book could be seen as criticism of a weak administration. I saw it as being more about *values*.

In his very recently published book, *Values Matter Most*, the noted speech writer (for Lyndon Johnson) and author, Ben Wattenberg, deals with values head-on. He does take the time to define the word, so maybe he knows what it means. His thesis is that the campaign theme in 1992, "It's the economy, stupid," probably was incorrect and now is definitely incorrect. Early in his book he says, "I think that *values* are our most potent *political* issue. I know that *values* are our most important *real* issue." This book is filled with charts and graphs put together by research assistants. Many of them are very instructive, and everyone who has any interest in politics should read this book. Contrary to Woodward's conclusions, Wattenberg sees Clinton as a potential champion of vigor and heroism, as embodied in the presidencies of Kennedy and Reagan.

Although *Values Matter Most* was a good book, it ran into trouble when he divided values into two categories: social issues and cultural issues. He defined them as social or "something for nothing." He suggests that everyone agrees on these. And then on cultural issues: "There is often no consensus about them, that is, Americans often don't agree about what to do about them." This thinking and all that flowed from it seemed like psychosociopolitical babble. Again, some very good concepts about the centrality of values lose focus and closure.

Social/Spiritual Writers

Stephen Covey bridges the gap from business and even politics to social/spiritual concerns. He takes theological and social truths and applies them to the world in which we live and does it in a very digestible fashion. His book, *The Seven Habits of Highly Effective People*, has had great staying power and has served as a foundation for a robust business, mostly consulting for industry. *Principle-Centered Leadership* (PCL) has been slightly less successful but no less illuminating. They make essential reading for anyone interested in VIRO.

In *Principle-Centered Leadership*, he says values are like maps, and they align correct principles with the territories he pictured. He suggests that maps become obsolete and that principles are like a compass, never obsolete. Principles are objective and external, he suggests; values are subjective and internal. The optimal state is to align personal values with correct principles. This is reasonable, but it is designed to place the emphasis on principles, what he is about to sell. I agree with his internal/external observation, and that leads me to have a bias to leading with values. Practically speaking, values, being internal, are what determine behavior and, therefore, untoward events. So, affecting values is essential for avoiding "accidents." Principles, being objective, stand outside this scheme and may define consequences of a given value set, but are without power to alter the underlying motive force, values.

Moving toward the spiritual climate, let's briefly review the popular book, *The Seven Spiritual Laws of Success* by Deepak Chopra. Prior to his Table of Contents, he quotes a passage from scripture, iterating my thesis that beliefs and values predict outcomes.

> You are what your deep, driving desire is.
> As your desire is, so is your will.
> As your will is, so is your deed.
> As your deed is, so is your destiny.
> —*Brihadaranyaka Upanishad IV.4.5*

In his Introduction, Chopra suggests that we all have seeds of divinity within us and need to nurture them. More directly, he says, "In reality, we are divinity in disguise, and the gods and goddesses in embryo that are contained within us seek to be fully materialized." Then he suggests what will be part of my theme: social laws are no less true than physical laws. He says, "The same laws that nature uses to create a forest, or a galaxy, or a star, or a human body can also bring about the fulfillment of our deepest desires." If one of the deepest desires (i.e., values) is safety, one just needs to learn the laws or principles governing safety and apply them in life, using them to produce the outcomes desired.

The book is filled with contemplation and wisdom that is organized into Chopra's seven laws. While all prove useful for reflection, two are particularly pertinent to the spiritual climate hospitable to VIRO. His Law of Intention And Desire counsels making a list of desires and then riveting on them...not allowing oneself to be distracted. If one values something, the ability to do it is enhanced. The Law of "Karma," or Cause and Effect, or "what we sow is what we reap," harkens back to the premise that beliefs and values predict outcomes. He writes: "In other words, most of us—even though we are infinite choice makers— have become bundles of conditioned reflexes that are constantly being triggered by people and circumstances into predictable outcomes of behavior." If the reflexes are conditioned correctly by establishing the right values, bad outcomes can be avoided.

Another law, The Law of "Dharma," or Purpose in Life, proclaims that people are not primarily human beings who occasionally have spiritual experiences but rather spiritual beings who are having occasional human experiences. The difference is profound. The intrinsic value of each human being and their deep seated moral compass and the values it leads to is the power behind VIRO.

Several times this book will refer to the book Richard Eyre wrote with his wife on teaching children values and on his book *Don't Just Do Something Sit There*, so let's not spend any time on his work here. In addition, other social/spiritual works will be referenced as it seems appropriate, not to mention some safety work, although that is less important because that is not where change is likely to come from.

THE CLIMATE SUMMARIZED

What kind of environment or climate awaits VIRO? The broad popularity of the diverse books mentioned before demonstrates what people in the USA are thinking and are interested in. In summary:

- A lot of business writers, politicians, and sociologists are talking about values...and people are listening.
- In a very integrated world, everything relates and fits together.
- Where there is segregation, it is unnatural and breeds disharmony.

- In spite of the interest in values, society doesn't have a good fix on how to teach them, particularly in organizations.
- People sense and feel the importance of values, but don't have a good way to measure them.
- Firmly held values that correlate with sound principles may be the strongest drivers in the universe.
- VIRO is at the interface of politics, social laws, business, and spiritual matters, and draws from all of them...and, yes, could contribute to all of them.

Before concluding, let me repackage the environmental characteristics discussed in the beginning of the chapter in light of the national political, business, and social climate. For clarity, they are separated into three categories.

Culture

Capacity to Understand. VIRO is intuitive; it harrows up the most basic instincts and inner "marching orders."

Culture Compatibility. This approach is culture friendly. By design, it accommodates *any* culture.

Driver Power. The drivers are one's values, so they are very powerful...maybe the most powerful driver in the universe.

Cost and Cost/Benefit

Need Must Exist. There is no "free lunch." So, we must be hungry if we're going to pay for it. With almost everyone talking about values, anything that focuses the discussion should be welcomed, if not embraced.

Cost Must Be Acceptable. With VIRO, you pay as you go. There is no up-front investment. Anyone can quit any time they want to without a termination penalty.

Perceived Benefits Must Exceed Cost. The return on investment will invariably be better than any other option, including doing nothing.

Administration

Explanation Must Be Clear. Read on. This isn't gene splicing.

We'll Need Champions. They are out there in great numbers, just waiting to be activated.

Preparation Is a Must. If this is an idea whose time is at hand, preparation will be easy. And if it slips up, inertia will take over.

3

LIMITATIONS OF TRADITIONAL AND BEHAVIOR-BASED SAFETY

"Good behavior is the last refuge of mediocrity"

Examining the theories and teachings of Edwards Deming, father of the quality revolution in Japan and now America, provides a treasure of ideas that have the potential to enrich American business and help us overcome some of the obstacles to renewing preeminence in the world economy. If asked to cite the characteristic of Japanese business practice that has helped them put America out of many industries we once dominated (and threaten us in many others), I would say, how they view errors or mistakes. The Japanese viewpoint emanates from their application of Dr. Deming's teachings.

The Japanese seek to *expose* mistakes and shortcomings so they can learn from them. Our American businessmen may say

PEANUTS   By Charles M. Schulz

PEANUTS reprinted by permission of United Feature Syndicate, Inc.

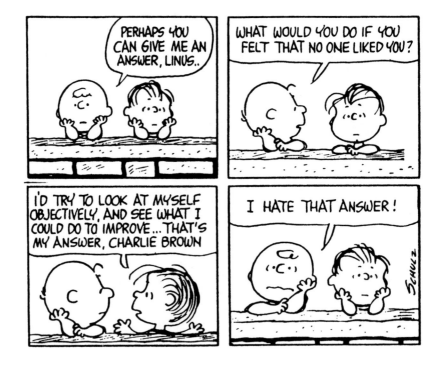

that they want to know about problems; however, the way they react to them sends a different message. America has such a history of bad things happening to those who deliver bad news that nobody wants to admit to a mistake or even to be the bearer of bad news. We generally take the ostrich approach to dealing with this. Even the companies with open door policies find that they are not working as designed. Or, maybe they *are* functioning as designed.

This two paragraph introduction is intended to justify what *may be* perceived as a trashing of some very good programs, practices, and initiatives. *It isn't intended to be that.* But if we cannot objectively examine what we are doing and accept the shortcomings and limitations of our historical approaches, we will never get any better.

In the first half of this century, the traditional approach to safety served America very well. This country made enormous strides and reduced accidents, fatalities, and probably numerous industrial illnesses ranging from cancer producing exposures to the less exotic, but nonetheless quality of life-degrading, musculoskeletal injuries. This progress has been less dramatic during the second half of the century. Some analyses would even suggest that America has made little or no progress in the last twenty years in spite of OSHA and all kinds of technology breakthroughs. Why is that?

This book will attempt to examine why the nation seems to be locked in the accident cycle by looking at the three main drivers of loss prevention (see page 34). These monopolize the attention of the decision makers; they must change their approach if we are going to achieve a breakthrough. The drivers are OSHA, the traditional approach to safety, and behavior-based safety. To begin, each of those approaches must be defined. Elements of each of them are found in almost every large organization in the country that has a proactive approach to controlling losses. OSHA

Limitations of OSHA, Traditional Safety, and Behavior-Based Safety

OSHA...

- has left a wake of frustrated administrators.
- reduction in accident rates is inconclusive.
- is government and government is rarely more effective than the private sector.
- has little impact on culture, which we all know is where the action is located.
- does not have a history of attracting the best safety and health people.
- uses the wrong driver—compliance statistics.
- mostly works through fear.

Traditional Safety...

- is based on a thirty to fifty year old model.
- is predominantly fear-driven.
- uses audits that mimic the regulatory process.
- is culture blind and insensitive.
- measurement is after-the-fact.
- punishes creativity.
- institutionalizes bad ideas.

Behavior-Based Safety...

- has become myopic.
- lets management off the hook.
- may eventually be seen as manipulative.
- stifles initiative in subtle ways.
- fails to deal with *real* root causes.
- provides a poor return on investment.
- lacks the power to be self-sustaining.

is not included in the chapter title for two reasons: (1) It has become a part of "traditional safety"; and (2) I didn't want to give it any more ink than I had to, particularly in large letters.

OSHA introduced a new era in safety when it was passed in 1970. No one who was involved in safety at that time can deny the enormous impact it had then and is still having on loss prevention. No matter how you look at it, it is regulation and requires compliance, or sanctions are applied.

Traditional safety was fostered by the most progressive and highly motivated organizations early this century. Insurance companies were the early drivers, and Liberty Mutual probably incubated more corporate safety directors than any other organization. DuPont was an early leader in safety and has a rich heritage that includes selling safety services. Whenever someone speaks of "world-class" safety efforts, DuPont will be mentioned; so perhaps it defines the traditional approach to safety better than any other organization and has made it work as well as anyone.

Behavior-based safety was introduced with that name early in the 1980s. Tom Krause deserves the right to be called the father of the concept, although many other proponents, such as Scott Geller and Michael Topf, deserve recognition. None of them are safety professionals and all of them have their origins in management and behavioral science, as is the case with most of the people "selling" this approach. There is a message in that and there isn't *much* wrong with it.

THE OCCUPATIONAL SAFETY AND HEALTH ACT OF 1970

Why shouldn't OSHA be used as a beacon to safety excellence? Here are some answers:

- The history of OSHA reducing accident rates is, at best, inconclusive.
- Those who have headed the program have left frustrated, not victorious.
- When is the last time government did something better than the private sector?
- We all know the secret to success in safety is culture/ behavior-based; OSHA has little impact in those areas.
- You get the best results with the best people. OSHA pays low and is not noted for high morale; hence, it doesn't attract many of the best people.
- Regulatory leadership is rewarded based on compliance statistics, not performance improvement—the old wrong driver problem.
- With almost no exceptions, OSHA works through fear.

In summary, America could do without OSHA with the possible exception of the OSHA Star Program, which has only attracted some of the best companies and has had very little impact on overall safety in the U.S. to date. I could come up with a few more but I'm probably "breaking through open doors" already. The rationale could be explained for each objection but it is not worth it. This text will spend more time explaining the thinking on traditional safety and behavior-based safety because these explanations will integrate with the purpose of this book, i.e., to suggest a better way to reduce losses using new methods but embodying the best of what we are presently using.

TRADITIONAL SAFETY

Where does the traditional approach fall short? What can we learn from our experience with traditional safety and then salvage what is best to develop an optimal model that will serve well into the future? Here is what is wrong:

- This approach is based on a model that is thirty to fifty years old.
- Like the regulatory solution, fear drives this approach too much; this is particularly prevalent in the programs that have often been touted as the most successful.
- When audits are used, they mimic the regulatory process and in this day and age, are largely obsolete.
- It is culture blind and insensitive.
- Measurement is after-the-fact.
- The process is designed to penalize creativity. (Old ideas die hard.)
- Some very bad ideas seem to be ubiquitous and have become institutionalized.

The Old Model Problem

Even if a Model T Ford is fitted with power windows and cruise control, it will still be an old car. When a mechanic tries to add air conditioning, he will find that the insulation, window glass, and engine just won't support it. The traditional safety model is so laden with dogma and procedures that confuse rather than edify, it would be better to start with a new format, delete the elements that don't fit, and modify those that do. For example, the safety policy statement that no one reads is anachronistic; yet the confined space entry procedure is essential. We need accident investigations, but we need to find root causes and address them. Almost none of the current approaches succeeds in doing that. We could go on for pages but that is not needed.

Fear Is One of the Primary Drivers

Fear is not good motivation, although it is tempting at times. I know I would have used fear gladly while riding to the airport this summer with my beloved three year old grandson and his mom, my daughter. Every five seconds for one hour, he asked for a drink using numerous approaches and various voice intonations. We were late to pick up his father and drinks were way off the highway. Only the love I have for my kind and understanding daughter and my respect for her authority in this matter kept me from using fear inappropriately.

I recognized that using fear would have been the easy way out. It doesn't require the patience, sensitivity, and compassion my good daughter was modeling for me. Those characteristics are not particularly prevalent in American management; perhaps that is why they like the fear approach. What are the bad side effects? You inspire anger and cynicism. You encourage people to cheat and lie to avoid punishment. Those on whom you used the tactic return the favor to others and the circle of fear expands. Management loses its credibility. Is that enough?

Recently, a company with "world-class" safety performance tried a period of amnesty for accident reporting, and their rates went up almost fifty percent. What is the message?

The Safety Audit

A recent project involved benchmarking the audit programs of some of the most prestigious companies in the country. All of them focused on compliance issues and had little or no emphasis on behaviors, much less attitudes, culture, values, or beliefs.

For the most part, they used evaluation techniques that measured and used grading rather than encouraging improvement. The emphasis is clearly on results rather than process.

One successful company, Proctor & Gamble, does a good job evaluating the behavioral aspects of their process in the audit/ evaluation. They have developed a sound safety culture, and if they decided to use a values-directed approach to safety, the transition would probably be swift, comfortable, and further solidify what they are doing. They are in a very small minority.

Culture Blindness and Insensitivity

Traditional safety programs rarely, if ever, consider culture. They are generally insensitive to special needs that people have. The few exceptions are the companies with enlightened employee assistance and wellness programs. But in most cases, even those fail to see the potential synergy with the loss prevention effort and take advantage of it. There is a lot of talk about the importance of company culture in business literature, but not much on how to measure or influence it. This book will talk about that and suggest not only that it can be done but how to do it. Culture and safety are rarely, if ever, mentioned in the same discussion. That's a problem.

The problem is illustrated by the fact that the same program, even when used with the same effort, works with varying degrees of effectiveness depending on where it is applied. The worst case is that one of the best programs (such as that sold by DuPont) doesn't work at all and is rejected by the potential hosting organization. Sometimes failure occurs and it is greeted with surprise and hollow excuses. The real reason for the failure is that the program was culture blind and insensitive from the beginning and couldn't overcome that deficiency.

Measurement When It Is Too Late

The three measurement techniques in common use (incidence rates, workers' compensation costs, and safety audits) that are part of traditional safety are all after the fact. Few people understand them well and reaction is spasmodic and almost never related back to process improvement. The traditional program is not bad at clean-up and reaction, but *prevention* is not the forté of the traditional approaches.

Creativity Is Penalized

The program is usually set in concrete and if suggestions are encouraged, they are more often than not ignored when offered. This breeds cynicism and estrangement. The exceptions are found in some self-directed work group plants that don't even have a safety professional. But the traditional program "gets them" in other ways. They often revert back to some of the most basic motivations, such as incentives and punishment.

Bad Ideas Are Ubiquitous and Thrive

Incentive programs are one example of a bad, ubiquitous idea. With very few exceptions, there is always a safety committee. The guideline should be to get rid of everything that starts with safety, including the safety professional. OOPS, that may be going too far. Seriously, you don't need many safety professionals when you know what you're doing. Even then, their roles need to be clearly defined and understood by everyone. They are helpmates only and are only accountable for the quality of their assistance, never the bottom line.

BEHAVIOR-BASED SAFETY

This is the best idea that safety has seen in the last thirty years and perhaps since we first decided that injuries were preventable. Having said that, what could possibly be wrong with it? Here it goes:

- Behavior-based safety has become myopic.
- When this is the keystone of a loss prevention effort, it can and has let management off the hook.
- It may not wear well with time if people see it as manipulative, which it actually is.
- It purports to encourage individual initiative but actually stifles or confines it to the parameters of the program, as dictated by the purveyors, whomever they are.
- It fails to deal with root causes in spite of claims to the contrary.
- It provides a poor return on investment.
- It probably does not have the power to become self-sustaining.

I feel badly about this list because I like the people who are selling these ideas and the concepts are better than the alternatives. There is a major role for this thinking in a VIRO company.

Myopia Is Endemic in Communities Practicing this Technique

Obviously, this refers to loss prevention myopia. If done as described in the literature, so much energy and effort is required to make this approach work, little time remains for anything else. How about off-the-job safety? What about wellness programs and employee assistance? How about engineering review of new construction? How about process safety? The future of the safety profession is in broadening, not shrinking, its scope. This point should be clear to anyone who has taken a close look at this process as described in the literature or witnessed it in action.

Management Gets a Free Ride

Everyone knows that management must have an important and ongoing role in the safety process, or behavior-based safety can't sustain its success. Behavior-based safety is largely sold on the basis that it is wound up and goes on forever. One of the recommended values is recognizing that safety efforts cannot end if the goal is to sustain an environment free of untoward events.

It Is Manipulative

People don't like to be manipulated. They react in many different ways, most of which are negative and destructive. They conveniently ignore messages; undermine the process; go underground if punished for being divisive; accidentally/on purpose misunderstand instructions, and so on.

When you tell people that you are going to provide soon, certain, and positive reinforcement for the behaviors you want to replicate and perpetuate; then you stipulate the behaviors and the methods to reward and teach them how to do it to each other; they may at first be amused. Soon they will feel that they are: (1) being treated like children; (2) being controlled in subtle ways; and (3) being used as instruments to control their friends and fellow workers.

It Stifles Individual Initiative

The road map to success in behavior-based safety is very explicit. It spells out, for example, who should be on the committee to determine what the critical behaviors are and how the committee should be run. It is all downright thought controlling. At a recent talk that I gave on VIRO, I complimented the behavior-based concept as I do in this book and still angered one of the conferees to the extent that he felt the need to stand and defend the concept. He had an almost religious zeal for the approach, seemingly to the exclusion of any other ideas. I find that a little scary because it smacks of mind control. There is never only one way to do things.

Most of us have seen the puzzle where you are asked to join nine dots in a square layout using four straight lines without lifting your pencil.

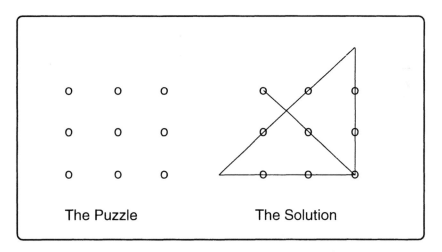

The Puzzle   The Solution

Most people don't solve the puzzle the first time they see it because they don't think to go outside the outer perimeter of the square formed by the dots. There are a lot of problems that cannot be solved unless people have open minds and feel free to go outside the square. In the case of complex, multidimensional endeavors such as loss prevention, we must go outside the "box." Going outside the box is not encouraged by the behavior-based concept purist.

Many companies say that their most important asset is their employees. In many cases, they fail to support that with their actions. This results in cynicism that corrodes the "team" approach and debilitates organizations. Behavior-based safety has the potential to severely limit the scope of employee contributions to their own safety and, in some cases, to reduce them to subhuman beings like the rats and monkeys used for experiments.

Root Causes Are Still Buried

Most other systems don't even claim to routinely find root causes. In spite of recognizing that it is hard to change attitudes and hence abandoning any such effort except by controlling behaviors, the behavior-based approach infers that behaviors are at the root of losses. I hope if you read this book carefully you will reject that out-of-hand. Addressing these attitudes coupled with claiming root cause identification is somewhat confusing, if not duplicitous.

Worse yet, when people find that they have been heading in the wrong direction and get incomplete results because they haven't found the root cause, they become frustrated. What follows is discouragement and demoralizing consequences. Whenever people treat symptoms instead of causes, that is what happens, whether with disease or accident reduction.

The Return on Investment Is Poor

In a day and age when competition is fierce and global, getting a good return on investment is more critical than ever. Installing the behavior-based process as the purist suggests is expensive. The cost for a plant runs into six figures and for a company of any size can easily run into seven figures, not including the down time for training and participation. It is by definition independent of other natural organization processes. This proves to be a significant obstacle for most organizations. A value of seeing safety as integral, a part of the whole (see Chapter 6), is recommended.

It Is Not Self-Sustaining and Renewing

Without a champion, this approach cannot sustain itself. It could actually sink under its own weight once funding is removed. There are ongoing costs for meetings, surveys, and retraining. Everyone knows that industry "gets what we pay for and inspect for and not what we pray for and expect." But if the cost is too high and the inspections too inefficient, it is only a matter of time before it will be cut out of the budget. Safety professionals should be empowering and working themselves out of a job. This does just the opposite. It requires constant maintenance and oversight. It bears some resemblance to the government that we are now trying to truncate with so much wailing and resistance.

SUMMARY

Okay, I have been harsh. But I think with some justification and more importantly, with a purpose. Before I state the purpose, let me mend some bridges.

I have the utmost respect for all who have toiled in the fields that direct themselves to conserving and protecting our human and physical resources. I believe that as a group, there are no finer people on the planet. Their cause is just and many of them could make more money in other work, but, their lives are laced with altruism. I have a great love for them and count myself fortunate to have labored with them and for them.

My net worth would be considerably less were it not for OSHA. I think that could be said for virtually everyone who has labored in the safety and health field over the last twenty-five years. Many entire businesses and even a few industries are an outgrowth of safety and health regulation. Most importantly, some people may be alive today who would not have been without OSHA. There is no way those results can be seen as all bad or even a bad idea. I just think there is a better idea.

The bedrock of processes that have produced dramatic decreases in accidents, injuries, and fatalities are identified as *traditional safety*. Without it, safety would still be in the dark ages. It must be seen as the base or foundation of any improvements that are made in the process. None of it can be rejected without being sure there is a better way. In most cases, that better way will evolve from what has already been conceived.

Behavior-based safety strikes near the heart of what causes almost all losses. The evolution of methods to enhance the understanding of behaviors and how to best measure, monitor, and control them may be the most important advance in loss prevention. The people who have pioneered behavior-based safety are, for the most part, not safety professionals but have been adopted into the family and are welcome additions. They have stimulated and enlightened safety professionals in ways they couldn't seem to find for themselves.

This text has tried to set up a straw man in which to poke holes without doing any serious damage. If the concept of a values-driven approach to conserving our physical and human resources is adopted, there may be a very important place in the process for OSHA, traditional safety, and behavior-based safety. The theory, if applied, serves as an antidote for the ills of each of those current drivers and lets them reach the full flower of their potential.

This flowering can occur, if everything done in the name of loss prevention is based on a set of values that everyone can support and is clearly derived from common beliefs. Then the obsession with outcomes diminishes until there is little or no concern for them except to the extent that they demonstrate and prove what is already known. Most people know that what you harvest is what you sow and "For where your treasure is, there will your heart be also." Everyone stops punishing each other for results and failures that may be outside their control and focuses on doing the right thing. Organizational neuroses naturally wither and fear is replaced with confidence and rationality. Audits start measuring process, which in this case are the values, culture, attitudes, and, yes, behaviors. They all come together in a one color harmonious quilt of loss prevention processes which is pleasing to the eye and functional.

Culture is an integral part of the approach and so can't be ignored. Everyone loves it because it honors their individuality. By its very nature, it always moves to the root causes. No one feels manipulated because they understand the basis for everything that is being done. Those who don't share the beliefs and would prefer to take more chances and maybe get injured need to become entrepreneurs or go to work for a company that sees workers' compensation costs as uncontrollable overhead.

Management takes on the role of organization patriarch and must set the example. They are intimately involved at every level of the process. Any failures of the processes are clearly laid at their doorstep, which is right where they belong.

Once the measurement is refined, there is little need for incidence rates or workers' compensation claims status reports. They only serve to comply with regulations and indicate what numbers to put on the check or in the budget. The OSHA log will show a lot of zeros. The few checks that are written will have small numbers on them. In two of the best plants that I have been in over the past several years, the management had very little interest in traditional measurement, although they were leaders in their groups and company and were held up on a pedestal when it came to safety performance. One was a saw mill, the other a paper mill, both operated by Weyerhaeuser in their southern operations. By the way, they were OSHA Star Plants. They were also the most profitable units in their business in the company. Do you find that surprising? I didn't.

But, they *couldn't* show what they believed in with regard to safety and how that translated into their safety values or how that related to their outstanding performance. They *could* demonstrate specific strategies, tactics, and plans that produced the great result. About all they needed to do was to think through and write down what they already knew and use it to remind themselves periodically what accounted for their achievements. That is why journal keeping is such a good idea. People tend to lose important ideas if they don't write them down.

I hope I have ended this very important chapter on a positive note. The past is indeed prologue. There is much in our past that is of great value and should not be lost. VIRO not only does not call for abandoning truths, it shines light on them and their proper utility.

4

RATIONALE FOR USING VIRO: A VALUE-DIRECTED LOSS PREVENTION PROCESS

"A man's actions are motion pictures of his beliefs"

VIRO is not just one more safety program —not a rehash of old ideas renamed, a new label based on a current fad, or just a restructuring. It is a completely new concept. VIRO is based on emerging but undeveloped thinking rumbling through society. The thought that values are important and that America has lost something over the last two or three decades as a result of a failure to hold onto that truth is being talked about every day in the media. Discussions of values surface in every venue from education to politics and even business.

VIRO will only confuse and irritate those seeking a quick fix or recipe for success that requires no thinking. It is much deeper than other safety programs that have come and gone. But it could stimulate the next wave in loss prevention and open incredible opportunities for safety professionals.

That sounds pretty heady and presumptuous, so I need to tell you where I think VIRO originates. I view myself simply as a collector, collator, and scribe. I have had the good fortune to associate with some of the finest and brightest people that anyone could imagine, both inside and outside of my profession. I discuss many of them in the acknowledgments. I have learned from all of them. My ecclesiastical leadership opportunities have provided me with wonderful associations. Some of the people I have served did not have prestigious educations but were full of the wisdom that can only come from living life with a purpose as a farmer, factory worker, engineer, or teacher. They taught me that much of my book knowledge was of little use when it came to things that really mattered.

Some of my church leaders have intellects far beyond mine and have been educated in the best institutions in the world. But they have offered me not only temporal but spiritual insights. They have stood on a higher "plane" than I have and, with great patience, nurtured my understanding of concepts that I feel have eternal significance and stem from laws that have been decreed before the world as we know it came to be.

As I have pondered what good I should be doing with my life, I have received constant and continual "gut" prompting on this subject that has enlightened my mind with ideas, the origin of which I really can't pinpoint. I know that they don't all come from my basic intellect or anything that I have been taught. Hence, I must conclude that I am just a messenger and collector of information. If this book and the ideas and activities that issue from it produce much good, I will find it difficult to claim credit.

Let's explore what I have called the keystone premise of VIRO: beliefs and values predict behaviors and outcomes. The complete linkage is shown below.

VIRO's Keystone Premise

BELIEFS ⇒ **VALUES** ⇒

CULTURE ⇒ **ATTITUDES** ⇒

BEHAVIORS ⇒ **ACTIONS** ⇒

INCIDENTS (OUTCOMES)

The terms *outcomes* and *incidents* are used interchangeably and there is a bias for the former term as the work progresses because it is a more generic term. It is more inclusive and more picturesque, a term that conjures up meanings that hopefully will lead to a better understanding of what VIRO is all about.

The words *descriptor/predictors* are used for the descriptions of levels of maturity in the maturity grid in Chapter 18 because they both describe where a particular organization is and predict the kind of results or outcomes it is likely to experience. The beliefs and values are *indicators* of what the descriptor/predictors should be and point to the outcomes that will be harvested.

During the past few years, one of my significant learning experiences came at an unexpected time in an unusual place from a surprising source. I listen to a lot of talk shows on radio and television and must confess that most of it is wasted time. One day I was listening to a radio call-in program as I accelerated up an on-ramp to the Adirondack Northway. I recall just where I was because what I heard struck me like a lightening bolt. The lady calling in said that she found it amazing that in our society so few people seem to be able to associate the bad things that happen to them with the wrong things that they do. She may have come closer to solving the great problems of the world than an elite team of social scientists and politicians could.

My wife likes to tell a story about my son that makes the same point in a real life situation. Most of us have experienced something like it and usually missed the real point. One day, Todd came home from school late and, being a stay-at-home mom, my wife was anxiously awaiting his arrival and concerned that he was late. She asked what had caused his tardiness. He quipped that his teacher had gotten him into trouble. She didn't just let it drop, but probed as to just how that had happened. He said that during class he was propelling small and harmless objects at friends who were returning the favor and she caught him. She then made him stay after school since she didn't see the activity as essential to the learning process.

Barbara spent some time asking questions like: "Didn't you know that what you were doing was against the class rules?" and "Is it any less wrong if you hadn't been caught?" and then "Can't you avoid detention by simply complying with class rules?" and lastly concluding for him, "Then if you get into trouble by breaking rules you know, you have no one to blame but yourself, right?" Recently, I had a discussion with him, at age 24, that showed the lesson had stuck. He had just had his auto insurance rates increased as a result of several moving violations. He understood that he had no one to blame but himself and that he needed to endure the consequences. He saw them as a natural and expected result of driving habits that he may just have learned from his chagrined father. (He is trying to adjust his behavior.)

Once we have grasped the correlation between our beliefs and values and where they naturally lead, it would be a shame not to apply that knowledge to increase the probability of achieving the most pleasing outcomes until we reach the blissful state of no displeasing outcomes. In the quality movement, the concept of continuous improvement is the cultural norm. How to achieve it is usually less well-defined. The safety movement has the answer and just needs to develop it, as this book should. Next, safety management can serve as a beacon to enhance how we manage everything. We will have new drivers that insure an optimized outcome with less effort.

What this values-driven approach can accomplish is described below.

Benefits Harvested From VIRO Application

- Facilitate Priority Setting
- Reduce Uncertainty
- Insure Buy-In
- Build Bridges to Other Disciplines (Unify Purpose)
- Substitute Action for Reaction
- Move to the Wellspring (Abandon Working Downstream)
- Elevate and Clarify the Process
- Quickly Reveal Ill-Conceived Approaches
- Provide a "Gut Check"
- Provide the Ultimate Yardstick

Facilitate Priority Setting

Currently, every organization concerned with safety has a different method of setting priorities, and they are almost as varied as the organizations themselves. Some write a lot of manuals and procedures. Others do a lot of training. Still others, who may not even be able to grasp what is being exposed here, use various misdirected incentive programs. If a group determines what is really important to them (deciding what they believe in as it relates to the protection of people, property, and the environment), and then establishes the values that result from those beliefs, it will start to become a lot clearer what is needed to be done and in what order. Priority setting becomes easier and makes more sense to everyone involved. Buy-in occurs naturally.

The usual pattern has been that companies become aware that some new regulation will be passed requiring them to tell employees or the outside world or regulators something about their exposures or loss history; then overnight safety becomes top priority and another chapter of the sporadic lurching to achieve loss prevention begins to be written. Or, a company has a serious fall, eye injury, or burn and can't think of anything but that subject for weeks or months. Focus moves to dealing with problems only reactively.

In more enlightened companies, the agenda for spending time and effort on safety is considered by representative teams with strong safety staff support. Those approaches make more sense than the knee jerk programming mentioned above. Unfortunately, these rarely seem to have a broad base of support that is sustained for more than a few months. Consequently, time dilutes them.

On a recent PBS special on the life of Edwards Deming, the CEO's of several of the largest U.S. companies were interviewed. They reflected on how the Japanese had embraced the Deming philosophy in the 1950s (after America had rejected it) and how it had been a major contributor to their industrial successes. Each had a slightly different perspective. However, they completely agreed on one thing: the emergence of Deming on this continent had dramatically changed the instructions they gave to their subordinates. They said that just a few years ago, they told everyone who worked for them, "Don't tell me about the process, just tell me what the bottom line is." They now tell everyone who works for them, "Don't tell me about the bottom line, just tell me about the process."

Unfortunately, the message hasn't traveled very far down into many organizations. Or at least it isn't being applied. Most managers have heard the language but don't really understand what it means or just can't shift gears. Or, maybe the culture of "what have you done for me lately" runs so deep in our society that a contrary message can't penetrate below the surface. I'm not sure what the answer is, but my experience tells me that we can "talk the talk but we can't walk the walk."

A group safety manager who was part of a massive company was recently interviewed. The company is talking about all the latest management crazes and the rhetoric is very impressive. This manager started the interview by telling of his first encounter with a new boss. The superior not only had safety but also quality responsibilities. During their first meeting, the leader, who had been heard verbalizing all the quality concepts exposed by Juran, Crosby, Deming, and others, started by asking his new report to show him the incidence rates and workers' compensation cost figures for the past two years. He then admonished him as the improvement was not to his satisfaction. There was no recognition that the safety practitioner had only been in the position for less than one year; that results were a line responsibility; that the records reflected programs and performance that had their origins six months to two years ago. Nor was there any discussion of process. The safety manager who was being interviewed was a former operations person who had learned safety on the job rather than in an academic environment. He had a good grasp of how accidents could be prevented, as well as how the work got done in this organization. He was clearly turned off, if not demoralized, by his new leader. To say he was confused would be an understatement. The interviewer could only recognize his dilemma, offer condolences, and change the subject.

Why had this happened and is it typical? To answer the second question first: It is all too typical. U.S. business has long focused on results over process, and despite reversals by management philosophers, there has been very little shift away from emphasis on the "bottom line" for the next quarter toward more focus on building a foundation for future successes.

The pregnant questions are "Why can't we move from knowledge to action?" and "How do we do it?" Let's start by musing about why we find ourselves in this apparent dichotomy. Change is hard. For about half a century, managers have been taught that the bottom line is all that matters. New thinking is seen as a fad that may soon go away. It is also harder and more threatening to think about rolling up the sleeves and really doing something, instead of just figuring out how to get the numbers to *look like* something *was* done.

The classic example of this is what happened at Chrysler Corporation. The company was run by accountants instead of engineers during a period that preceded their near demise. The accountants realized that they had a lot of assets on the books at depreciated values, which in the real world had actually appreciated and could be sold for a lot more than the book value; these included factories, machinery, and property. For a while they just sold little things to make up the difference between earning projections and reality. They became addicted to the process as it was much easier in the short term than engineering, manufacturing, marketing, and selling the best cars. The problem was that when the property ran out and the foundation of good engineering and cars was crumbling, the company could only be saved by a benevolent government and some courageous executives. Is this any way to run a car company or, for that matter, any organization?

Our whole culture is screaming through the media that instant gratification is the way to go and that happiness is the direct result of pleasure. Few people stop to realize that the opposite is true and that they are being led astray. We know that the pleasure of a trip is in the trip itself and not in the arrival. We know that if we reflect on our lives, the things we are most gratified by are usually the hardest things we have had to do, not the easiest. We know that the things that give us the most instant gratification, like overeating and pleasing our senses in other ways, always demand a price that usually makes us unhappy. Why can't we remember that when we see the ice cream advertisement or the suggestive movie that encourages promiscuity or even deviant behavior? Perhaps because we forget what we believe in, what we really value, what is in fact important to us. We lose sight of the target.

If we are target shooting, would we prefer to be blindfolded, spun around a few times, and given ten shots? Or would we rather be placed in a dark room with the target illuminated, no obstructions, and a single try to hit the target? The answer is obvious. If we start by riveting on what we believe in and hence what we value, we've taken off the blindfold and turned off the lights so we can focus on the illuminated target and avoid distractions. The chance of hitting the target then is far greater. We substitute a more certain process for one that is random. Starting with a foundation of beliefs and values—no matter what the subject—facilitates priority setting.

Reduce Uncertainty

Very few people like uncertainty, and they usually try to reduce it. Senior executives and other leaders have styles that vary widely, and it is hard to isolate particular characteristics that are invariably associated with success, however measured. They almost all have short attention spans and don't like surprises, even if the surprises are good ones. In one way they are like small children and in the other way, just the opposite.

If we manage from a foundation of values, surprises are anomalies. The requirement for constant attention is eliminated. Tedium is avoided.

One of Peter Drucker's sayings is that: "The only things that evolve by themselves in an organization are disorder, friction, and malperformance." This appears to be one of the laws governing the cosmos. In physics, the second law of thermodynamics states that disorder (entropy) in the universe is always increasing. Work (in physics and business) is required to maintain order in a given system. So management must use methods that encourage a process that predicts order, harmony, and exquisite performance if they are going to compete effectively. VIRO is such a method.

Ensure Buy-In

Most people who have been involved in any type of management or group process recognize the need to get people on board. If it wasn't important years ago when autocratic management was more stylish, it sure is now. Telling people to do something may have worked in bygone times but doesn't have a very high success rate in our current society.

Whenever something needs to be done, one of the first questions is, "How do we get buy-in?" There are usually many answers, most of which involve coercion. We try to show people what is in it for them or give them rewards for doing what we want them to do. These techniques are not necessarily bad and in the short run may be the most effective and expedient approaches. Taking a longer view, there are several problems. What happens when the budget is cut and the rewards are diminished or even deleted? What happens when the rewards don't meet expectations, as they rarely do?

Safety consultant Jack Gausch likes to tell the story of how, when he was with Merck & Company, he advocated rewarding his facility employees with pens for a noteworthy safety achievement. He had a thousand employees, so he awarded one thousand pens. There was a ten percent pen failure rate, so one hundred were returned to him. He ordered 100 more. As could have been predicted, ten of those came back. He got wise the next time and ordered a few extras. A few weeks later, people started to stream in inquiring about refills for the pens. Not being in the pen refill business, he was not always as accommodating as he might have been. Was this a good idea for him, the employees, the safety effort, or the efficiency of his organization? It's not clear. But it sure dampened his enthusiasm for incentive programs and he probably never again considered pens.

Showing people what is in it for them can be very effective, particularly in a society that is becoming more self-centered each day. Could it be that society is driving this move to self by the methods of achieving buy-in? If so, it may be creating a monster that it will soon lose its capacity to feed or house. Is this scenario familiar to any parents you know...maybe even someone you know very well?

Is there a better way to get "buy-in" and enhance certainty of expectations? Perhaps. It is not only better but doesn't require a lot of baggage that may sink the entire effort some time in the future. If a link can be established between outcomes that are desired and what a person values, their interest will invariably be attracted. If it can be done on a group basis, there really is a reservoir of power.

Let's look at a family that has just finished dinner. The four children of both genders and different ages are assigned clean-up. They can't decide who will do what based on history, age, other obligations, and so many other variables that a master facilitator/negotiator would throw up their hands. The parents have numerous choices of how they can handle the situation. Let's examine some of them and start with the position that there will be no dessert for anyone unless clean-up is accomplished. Dessert is not a reward but a normal part of the meal that has conditions associated with it. Some of the choices are:

1. The parents can dictate who will do what based on their idea of fairness and stipulate that those who comply get dessert and those who do not must go to their rooms and do their homework.

2. The children can be told that they must work out who does what and the parents can go and read and return at a set time. If clean-up is accomplished, they have dessert together. If the work is not done, no children get dessert.

3. The parents can teach the value of teamwork and the concept that "many hands make light work." They can build on the belief that the family is a microcosm of the world and lessons learned in the home about cooperation, sisterhood,

brotherhood, and serving each other will make life outside it easier and more enriching. They can then leave the children to work out how they will clean-up and accomplish the other things they have to do together and allow them to experience the natural consequences of their behavior and encourage them to see the results as those natural consequences.

Obviously, the third choice requires preparation and possibly the most time. But when done, it offers the seeds of understanding the meaning of brotherhood and sisterhood and avoided bickering over details. All the children come to understand outcomes as a natural result of their beliefs and values, and they enjoy a family culture of pleasing and harmonious behaviors. They become self-reliant and eventually extend the knowledge gained to other more complex environments.

Build Bridges to Other Disciplines (Unify Purpose)

In *Reengineering the Corporation*, Hammer and Champy ridicule what they call the Humpty Dumpty School of Organizational Management. They describe it as the breaking up of tasks into little pieces and record a legacy of inefficient organizations due to these practices. Implementation of a focus on values will not only discourage the corrosive practices described, but also provide bridges to other disciplines and encourage unity of purpose.

A study of the quality process demonstrates that the methods were either parallel to or identical to those applied for accident reduction. Is it any wonder since both disciplines are designed to produce fewer undesired outcomes? Quality deals with off-specification product; safety deals with accidents that result in injury. Both are unwanted events. Chapter 19 suggests some of hundreds of concepts that apply equally to both disciplines. I used them in a seminar entitled "TQM and Safety Parallels."

As an outgrowth of my quality reading and seminar teaching, I think I have stumbled over two other truths. The *first* is that the methods held in common in the quality and safety processes can be used in almost every other discipline in a complex enterprise, whether marketing, selling, manufacturing, materials handling, engineering, or human resources management. The *second* is that VIRO may be the elusive bridge that could allow an organization to unite the methods and produce synergy between disciplines in a non-threatening way that avoids turf battles and squabbling.

Safety is probably the least threatening discipline in an organization. No one wants to see anyone hurt or property damaged. Management and workers unite on their objectives. They only disagree on methods. The methods applied to produce a product free of defects such as "do it right the first time" and "success is rooted in relentlessly pursuing root causes" apply in everything we do in business and life. Why can't we see the similarities and use them to reduce redundancy? The answer is unclear to me; no one else seems to have found it yet either.

If it can be demonstrated that the best way to manage loss prevention is to base actions on sound beliefs and the values that are a natural outgrowth of those beliefs, the next step could be to show that the same methods will work to manage other facets of an enterprise. The beliefs and values will serve as a translator or common language that will result in unity of purpose.

The "management by walking around" that Peters suggests is a hallmark of excellent organizations and will be encouraged; even if people can't exactly understand what others do, they can share beliefs, values, and aspirations. The common threads knit together relationships that can help a company grow and prosper, which at its best is an extended family. The Japanese certainly accomplish this—probably to a pathologic extreme.

Employees already practice some of this extended family building. They are intuitively aware of the benefits, so they will readily embrace the concept of values-driven safety. Management has effectively kept them from it up to this point and just needs to get out of the way. As Drucker said, "So much of what we call management consists in making it difficult for people to work." We could add to that "or discover better methods," which is what a values-driven process is, a better method.

Implementation should be both bottom up and top down. The former will be easy; the latter very difficult. Employees will love this kind of participation that deletes duplicity, deceit, and deception, often the three D's of current safety programs. All but the most self-assured and enlightened manager will have a hard time mustering the courage to abandon the incomplete processes of the past and build a new foundation. One of the beauties of VIRO is that "you don't have to throw out the baby with the bath water." The company can keep everything good that they are doing and allow the wasteful to wither and die. VIRO represents higher, unifying law.

When other disciplines see how well VIRO works, they will want to embrace a similar approach and the natural result will be values that bridge other disciplines and unify purposes.

Substitute Action for Reaction

The current paradigm in safety is inspection and reaction. Whether with OSHA compliance, accident investigation or an audit program, we inspect and then react. The process is analogous to a dog chasing its tail. Using VIRO results in reaction being replaced by action. Rather than waiting for an untoward event and then reacting, one should decide what to believe in to encourage the appropriate values. These values produce a culture that spawns attitudes leading to behaviors that predict the desired outcomes. Adjust beliefs and values to achieve long term solutions. Part of every management meeting will be devoted to measuring and fine tuning the company beliefs/values/culture to insure the outcomes they desire. All untoward events will be traced back to belief/value/culture deficiencies, and they will be adjusted, installed, and perpetuated. Action replaces reaction.

Move to the Wellspring (Abandon Working Downstream)

Following this model, VIRO moves action to the wellspring instead of working downstream. Shaping a newly trickling stream requires far less effort and is more effective than high energy attempts to control a downstream torrent.

Current approaches are almost guaranteed to produce what is called the accident cycle. Loss history goes up; people react and it goes down; people relax and it goes up again, and so on.

The behavior-based approach is marketed as one way to stop the accident cycle. It is probably better than anything used to date but has the flaws discussed in the last chapter. It works on behaviors that are upstream of accidents but well below attitudes, cultures, values, and beliefs. It will work under certain circumstances but is more fragile than VIRO and would be most dependable *within* a values-based context.

The traditional approach relies on adjustments to the particular program that broke down and resulted in the loss. Society is constantly plugging holes in the dike instead of building new dikes out of improved materials.

The working model of VIRO exists in a few organizations but is not well-documented. They know what they believe in and some have even defined a set of principles by which they live. These principles should be called values and be defined by the beliefs that spawn them. Employees are the ones that get hurt and make the errors that result in losses. They understand and relate better to beliefs and values than to mission statements and principles. Some may say this is just semantics. But words matter, and the right terminology is essential so all will embrace VIRO.

A friend of mine, Joan Brayton, was aware of my book writing and ideas. She asked how you would keep people out of the organization who didn't share the same beliefs and values the organization adopts. My answer is one of the reasons this concept is so powerful. There will be a natural selection process. There will be no need for any type of discrimination. People who don't hold the company's stated beliefs and values won't want to be an employee any more than someone will join a church, civic group, or special interest organization if they don't value the same things that organization values. Moving to the wellspring instead of working downstream makes employee selection and everything else easier. Basically, the company will *automatically* "do it right the first time."

Elevate and Clarify the Process

When a person defines their beliefs and values that support those beliefs, the process is moved to a higher plane and the purpose is more clearly defined. Individuals and organizations who want to make best use of their time recognize that there is a hierarchy of purposes. Activities always expand to fill the available time. Unfortunately, they don't always fill the time in their order of importance to employees or the organizations. Books have been written on the importance of prioritizing. Prioritizing itself, though, is exceedingly difficult as so many unique conditions and variables come into play.

The best way to set priorities is to come into intimate touch with what is important to oneself or the institution or organization served and then base priorities on what has the highest value. If two coins lie on a table and only one can be picked up, a person would obviously select the one of greatest value. Why not use that as a simile for making choices for loss prevention? Do that which has the most value. It can't be done unless there is a clear and certain knowledge of what *is* of greatest value. By grounding the safety and health process on beliefs and values, priority setting becomes easier and is far more likely to be ego-syntonic.

In most cases, employees set priorities based on what they *think* management wants them to do. That is dangerous for both the employers and their leaders. Dow's Law states: "In a hierarchical organization, the higher the level, the greater the confusion." The law is not absolutely true, but many of us have had experiences that seem to prove the law. The best way for leadership and subordinates to insure that everyone is pulling in the same direction is to all be unified on priorities. That will best be accomplished if all have the same purpose. Purpose will be most clearly defined if beliefs and values are held in common.

In a recent *Root Cause Network* newsletter, the masthead read, "Safety Should *NOT* Be Your Top Priority." The point the author made in the article was that priorities change but values do not, and hence safety should be a value, not a priority. It is agreed that people should always place a high value on preventing losses. If a foundation of values is set for the loss prevention process and referred to constantly, people will sense when they have changed or modified and respond logically. The entire process is moved to a higher plane and the purpose is automatically clearly defined. However, the values that dictate how safety is done may change.

Quickly Reveal Ill-Conceived Approaches

As a natural by-product of having clear and unified purposes, diversionary and ill-conceived approaches are easily discerned, avoiding time wasting activities. Encouraging suggestions is almost always healthy. The difficulty comes in evaluating which ideas are good and which are not.

When it comes to safety, everyone seems to have an opinion. The same people who would be reluctant to give an accountant, engineer, doctor, or lawyer advice will not think twice about telling a certified safety professional how to reduce accidents. While not all bad, it sure creates confusion. The best way to avoid the confusion is not to automatically reject ideas but to have some way to test them quickly and render an opinion that has a high likelihood of being understood and accepted by the suggestor. To do otherwise will stifle participation, which is fundamental to preventing the wrong outcomes.

At a plant staff meeting, we were discussing the lift truck driver training requirements. The plant manager told me to assign the training chore to the safety engineer who worked for me. I suggested that he could do it but that if he did, it would occupy about ten percent of his time for the next year. We had several hundred lift truck drivers at the plant. I further indicated that if we assigned him chores like this training nine more times, we wouldn't have a plant safety engineer any more. He got the point and placed the responsibility where it belonged, with the materials handling manager. I had a narrow escape that could have been avoided if one of the safety values of the organization was to empower others (see Chapter 14). It would provide a litmus test for the plant manager's idea. There is no need for whining or looking like a malcontent or just appearing lazy. It is clear that training lift truck drivers does not meet the spirit or letter of what the organization thinks the safety professional should be doing to fulfill their role in the group.

Incentive programs, too, are easily identified as distractions if the group accepts the value that with loss prevention there is no end (see Chapter 7). Awards programs can reduce workers' compensation costs and even accident rates, but everyone knows that the effects are transitory and rarely work on the root causes of undesired outcomes. In fact, they often serve to mask the real causes due to cover-up activities prompted by peer pressure and appeals to greed and materialism. Not many people would see them as virtues.

On the other hand, valuable new initiatives can be quickly seen as virtuous and valid. For example, the value suggested in Chapter 8, "First It is a People Business; Things are a Distant Second," justifies the efficacy of behavior-based approaches. And those approaches are seen in perspective—they are probably not an end in themselves but a means to some ends and require a values-driven context for longevity.

If specific elements of the traditional approaches to safety are examined in light of one's beliefs and values, it will be easier to determine which of them are central to the quest for avoiding loss, whether of human or physical assets. Again, VIRO will help you discern diversionary and ill-conceived approaches as well as bring good ideas into better focus and implement them in concert with your overall plan.

Provide a "Gut Check"

VIRO can also serve as a vehicle for a periodic organization "gut check." As times change and resources are adjusted, people must periodically check their activities in light of new situations/circumstances. Accountants call it zero-based budgeting; management calls it reengineering. Failure to make mid-course corrections may mean missing the target or making costly redirections later. Obviously, making frequent small corrections is far better than making dramatic changes that cause disruptions and are inevitably costly. Many companies slept through the economic changes in the late 1980s and were forced to make dramatic cut-backs in the 1990s, resulting in trauma and social problems from tension and uncertainty. Wouldn't it have been far better to remain alert to shifting circumstances and make many small and planned changes?

IBM had a very recognizable culture that served it well for decades. Many of the elements of that culture would serve it well today. Other elements were almost fatal. The company was self-satisfied and inbred. Their reaction time was slow and they couldn't see the importance of quick reaction and outside influence. When they did, it had to be done at the very top of the organization, with the accompanying trauma. Traumatic changes may have saved the company but did not come without a price. If they had written down and regularly tested what they were doing against the beliefs and values that spawned their culture and adjusted the values against the current business environment, they could have made adjustments earlier and avoided certain material and human costs. The lesson here is that a company can't test what it is doing against its beliefs if it doesn't know what it believes.

On a broader note, let's look at one of our greatest current social problems and see how a VIRO approach could help. The nation's leaders have finally discovered that the breakdown of the traditional family unit is at or near the origin of most of our social problems. Now we are discovering that out of wedlock births strongly correlate with dysfunctional family units. It may seem that we have known these things forever, but remember that few talked of the centrality of the family until Vice President

Dan Quayle raised this issue in the 1992 presidential campaign, although in a manner probably not in his best interest.

Why do these discoveries come so late that the country has to fumble for solutions like cutting off welfare to dependent children of unwed mothers? Our social scenery would be much different had the country been more vigilant at promoting and living by the high moral standards valued in Jewish, Christian, and other traditions held by most Americans. On an ongoing basis, we could have been testing how we were doing against those values. Seeing the increase in illegitimacy long ago correlating with the wrong outcomes of increased crime, decreased learning in the schools, and other social ills, we could have acted aggressively and swiftly to avoid the current problems.

Over a decade ago, a feminist single parent professor at Princeton University was offended by a comment she felt was depreciating to single parent families headed by women. The person who made the comment had suggested that many of our societal ills correlated with fatherless families. The professor decided that she would set out to prove the allegation wrong. She embarked on a study that consumed many of her most productive years and proved the comment was more correct than the speaker would have ever guessed. She has shown direct correlation between school drop-outs, incarcerated teens and post-teens, and illegitimacy with fatherless homes. She has defended her results from confounders of economic bias and other factors, virtually all of which she considered in her comprehensive studies. Didn't we know the answer before the study? Did we have to do the *accident investigation* to know that fatherless homes were a problem?

Once people know what their values are, they can, on a programmed basis, measure how they are doing against them, and if the values are sound, they will see problems or bad outcomes coming long before they are overwhelming. So, a

"Well, here's your problem, Mr. Schueler."

values-driven approach offers a periodic "gut check." The mechanics of using it as such will follow in due course.

Provide the Ultimate Yardstick

Lastly, and perhaps most importantly, with VIRO, the ultimate yardstick is provided. Every business leader wants and needs better measurement. Not all executives subscribe to the philosophies of Peter Drucker, but few would argue with his thesis that "if you can't measure, you can't manage." Better measurement comes in two forms. The first is improved accuracy and precision (i.e. reproducibility); the second is relevancy and timeliness. The best case is totally accurate, precise (reproducible), and predictive (relevant) measures so that adjustments can be made to achieve targets and objectives.

Over a decade ago, I had a discussion that I have pondered many times since with my boss and mentor, Ted Mullins, the vice president for human resources at Chesebrough-Ponds, Inc. I casually suggested that safety may be the best early warning signal for business vitality in any company. In retrospect, we had a revolutionary idea, but unfortunately neither of us seriously considered it.

If beliefs and values predict cultures, attitudes, behaviors and hence outcomes, all we need to do is determine the values that correlate with the outcomes we desire and measure them to know what our accident rates, workers' compensation costs, or safety and health audits will be. We'll be able to accurately predict accident rates three months to three years before we have results. If the thesis is correct that safety cultures and performance parallel business cultures and results—that outcomes in the former can predict results in the latter—when we devise a reproducible, correlated with a safety and health measurement system, we will have the ability to predict business success as well. The correlation between organizational values and safety performance will be easier to demonstrate than the relationship between safety performance and profitability, although some day that will be able to be achieved.

At the very least, if it can be demonstrated how organizational values can be measured and then tied to injury incidence rates and costs, the process can be fine tuned and used to predict performance. Chapter 18 provides a system using maturity grids based on suggested values and assessments of superiors by subordinates to take the value "temperature" of an organization.

If it can be demonstrated that values predict outcomes, measuring values with maturity grids (see Chapter 18) provides the ultimate yardstick: the maturity grids that gauge organizational values. Most people are familiar with the exit interviews taken at polling sites that now predict election outcomes within a couple of percentage points based on a fraction of one percent of the electorate. If this kind of surveying technology can be developed, why can't beliefs and values be surveyed to predict accident rates and profitability?

I hope that in the above discussion I have persuaded the reader that VIRO may be the best way to conserve/optimize physical and human resources. The theory must be tested. Can you afford not to do it? Your business adversaries may try it and gain a competitive edge.

SECTION II
SUGGESTED VALUES AND RATIONALE

OPTIONS AND REFLECTIONS

5

DO IT FOR THE RIGHT REASONS

"Without a good reason for doing a thing, we have a fine reason for leaving it alone."

This is the first of ten chapters that discuss values suggested to predict a loss resistant environment.

INTRODUCTION

When I got irritated at my children, I had several homilies loaded with clichés that I used to fire at them that included such wisdom as "If you refuse to learn from your mistakes, you are condemned to relive your failures," or the familiar "Those who fail to plan, plan to fail." When I delivered my sermonettes, I had the feeling that they weren't listening.

Lo and behold, now that my children are grown and have their own children and have little to fear from me, some startling facts have emerged in family discussions. They not only didn't listen very well but used to mimic me and make fun of me behind my back. Part of my motivation was indeed to help them avoid negative experiences. But, in reflection, more of my motivation was to vent my own frustration. I guess I succeeded in doing the latter but doubt I did much of the former. Actually, the horror of all this is that I probably encouraged resistance to the wisdom I thought I was imparting by rubbing their noses in it. At least three of them, and possibly the fourth, inherited my stubbornness. The good news is that the supposed wisdom may have often been flawed or inaccurate.

I'm not alone in having misguided motives. In her book *The Fourth Instinct: The Call of the Soul*, the lecturer and television personality, Arianna Huffington, relates numerous experiences with celebrities who are wealthy or famous or both. She discusses the paradox that almost all of them have so much and feel so empty. Could it be that their accomplishments were achieved for the wrong reason?

What does this confusion have to do with the suggested first value to predict a loss resistant environment? If you do something for the wrong reason, you won't derive much satisfaction from your efforts and may even sow the seeds of your own destruction.

In business, with a pure bottom line philosophy, how or even why something is done doesn't matter. Only the result matters. Our nation's industries have long held the bottom line philosophy. Recently, the nation has become aware that how things are done predicts the nature and the quality of the product. If they focus on the end result, they get the most of it most of the time, but along the way they often lose things like quality, credibility, labor relations, and overall organization health. Reconstruction can be torturous. Many companies are in the reconstruction mode now. Some others have disappeared with nothing left to reconstruct. Witness the U.S. television manufacturing companies. There are none. Could it be that their failure to take a long view or "do it for the right reason" contributed to their demise?

Early in my career with Chesebrough-Ponds, which lasted thirteen years and ended with the acquisition of the $3 billion plus company by Unilever, I had an experience that has stuck with me over the years. It will help further set the stage for what I want to communicate in this chapter.

Chesebrough-Ponds' CEO, Ralph E. Ward, was a uniquely capable executive in many ways. He took a modest sized health and beauty company and through acquisitions such as Bass shoe, Stauffer chemicals, Prince tennis racquets, Ragu spaghetti sauce, Health-Tex children's clothing, Polymer Corporation, and Prince Matchabelli, built a successful and complex industrial giant. He was not a college graduate and had started with the company as a clerk. He had an almost mythical image in the organization and was viewed as somewhat enigmatic, even by many officers of the company.

The petroleum jelly processing plant in Perth Amboy, New Jersey, where Ralph started his career had worked for over a million man-hours without a lost time accident. They were having a presentation and luncheon. Both he and I were invited. With numerous locations all over the world, he often politely declined to attend and delegated such responsibilities. This one was where he had started with the company and many of the workers were old friends. He accepted the invitation.

I found myself seated across from him at lunch. He knew who I was, but we had to that point had no personal contact. Our eyes met and he obviously felt the need to say something. He asked, "How are we doing on OSHA compliance?" In everyone's life there are defining moments. I sensed that I may be in the middle of one. Should I tell him what was safe; what I think he expected to hear, or should I tell it like it was.

Being a confirmed risk taker, I opted for baring my soul. I responded that I wasn't real sure about that since I didn't spend much time on it.

He became more alert and riveted his eyes on me and suggested, "I thought that was what we hired you for." I knew that there was no easy way out at that point and that I needed to be at my best if my young family was going to continue to be warm and well fed and if those children would ever realize our desire for them to be college educated. My life quickly flashed before me, and I delivered my short speech that, on reflection, I believe I had prepared in quiet meditation many times as I contemplated the opportunity to actually address our CEO. The window of opportunity that I had hoped for was opening. I needed to jump through.

I said something like, "I was operating under the impression that I was hired to support the primary objectives of the corporation which I consider to be your concern. I understand those to be providing a good return on investment for the stockholders by paying solid and increasing dividends on the stock they own. My role in helping you do that, as I see it, is to be vigilant in recognizing areas where we are incurring unnecessary losses or where loss potential exists. I then communicate that to appropriate operations managers and provide suggested solutions, preventive measures, and counsel and assist them with the solutions we devise together. OSHA is a small part of that but, quite frankly, more of a diversion than a solution, so I don't spend much time on it. We have attended to the ten most frequently cited items and I don't think you or the corporation will ever be embarrassed by an inspection."

He looked down a little bit, then looked up. I started to have a good feeling that I can still remember and some of my anxiety left. He smiled slightly and said, "I like that, keep doing it."

As a footnote to this brief exchange, over the next ten years, we had numerous OSHA inspections. The total fines paid in this corporation of about one hundred locations and 40,000 employees were under $5,000 at the time I left. About five years after this discussion, I was summoned to the office of the Chairman and informed that I would report directly to him. I'll discuss why and how that happened later in this chapter and elaborate on the experience in Chapters 9 and 11. I doubt it would have happened without our luncheon dialogue.

Here are the wrong reasons to do safety:

The Wrong Reasons to Do Safety

- OSHA—Compliance isn't a good reason to do anything.
- Accident Statistics—They are inaccurate, negative, and confusing.
- Costs/Profits—They distort some important issues and mute others.
- Management Directives—No one really listens.
- Inspection and Audits—They often mirror compliance.

OSHA AS A DRIVER

My opinion about OSHA has changed little since my statements to the chairman, but my reasons have deepened. Executives in an organization look at regulations as annoying and undesirable obstacles to their success. They see compliance as depleting the human and financial resources that they could otherwise use to accomplish their primary objectives. It doesn't matter if that perception is accurate because most people recognize that perception is all that matters. Reality is academically interesting but rarely much of a factor in what people do unless it coincides with their perceptions. By the way, I believe executive perceptions of OSHA happen to be close to reality.

So, loss prevention to comply with regulations rarely gets management support, at least not without dragging them into the process screaming and yelling. Those who champion compliance will always be viewed with a jaundiced eye and may even be seen as pariahs in the business community. Is that any way to get started on building credibility, which is the foundation of any wholesome and productive relationship?

Management's perspective is not the only concern when contemplating compliance. Let's reflect on how workers view regulation. Their view is far different than management's and not nearly as one-dimensional. In a way, they see regulation as good because it *can* protect them. They can use it to leverage other things, using it as a bargaining chip in negotiations. It may also reduce the pace of work and lighten their load. On the other hand, some will see it as a nuisance. It may require them to do or wear things that are uncomfortable. So even for the workers, OSHA is at best a mixed bag.

The net effect is that OSHA has many negatives and a few divisive positives. Employees feel that management is the adversary and that they really didn't want to do the right thing but had to be forced to do it by the government. In *The Seven Habits of Highly Effective People*, Stephen Covey lists one of the habits as Think Win/Win. As a driver for conservation of physical and human resources, using OSHA compliance is a Lose/Lose option with a very negative flavor.

ACCIDENT STATISTICS AS A DRIVER

How about accident statistics? The problems are summarized below, followed by explanations.

Problems with Accident Statistics as a Driver

1. They are rarely accurate.
2. They are negative.
3. They only represent the tip of the iceberg.
4. The exercises that produce them (the investigations) are usually poorly done by the wrong people at the wrong time and produce an inaccurate result that is regarded with cynicism by all the employees and most of knowledgeable management.
5. Few people actually understand them as they are commonly displayed.
6. The consequences associated with them range from none or occasionally negative for the workers to jeopardizing career paths for managers in some companies. Hence, reaction runs from ambivalence to fear. Neither a very good driver.
7. The nomenclature is inconsistent.
8. Efforts for unique displays and interpretations tend to confuse rather than edify.

---

1. Accuracy. In the discussion of traditional safety in Chapter 3, fear is cited as a major driver and offers the amnesty example that increased rates fifty percent in one large company. Some landmark OSHA citations issued to Fortune 50 companies have shown seriously flawed recordkeeping at those companies. After one acquisition at Chesebrough-Ponds, which came at about the time of several of the aforementioned citations and fines, they examined the rigor of the acquisitions counting and doubled the number of recordable injuries reported.

If one divides severity by frequency and then divides the days lost by the total injuries, they get [days lost/lost time injury]. Since severity is [days lost/time frame] and frequency is [lost time injuries/time frame], they cancel the time frame (whatever it is) and are left with [days lost/lost time injury]. Then dividing the total days by the total injuries reveals how many days are lost in an average lost time injury. This exercise can unmask creative recordkeeping prompted by ethically disoriented people who will always be with us.

Typically, days lost per lost time injury should fall within the range of ten and one hundred. Anything less than ten indicates experiencing and counting many very minor injuries, some of which probably should have been able to work. Over one hundred starts to cast doubt on whether or not all minor injuries are being counted. Before OSHA recordkeeping, there were numbers over four thousand. That suggests only two possibilities. Either the company is not experiencing minor lost time accidents or is finding ways to hide them. You can decide what I concluded.

OSHA recordkeeping, which has essentially replaced the ANSI injury reporting system, recognized the probable burying of many significant injuries by returning injured employees to the infirmary or make-work jobs. The OSHA system calls for counting any light duty or alternate work as lost days and mandates recording any injury that required two doctor visits. The objective was noble. The result was a two edged sword that still confuses and misrepresents. Figures don't lie but liars figure. So, accident statistics are at best confusing and don't make very good drivers, even if they are accurate, which is rare.

Why does the OSHA system fail? In many cases, it suggests just the opposite of what is happening. For example, the caring and conservative company will provide excellent medical services and good follow-up. Their reward is to be penalized by having more recordables. Also, providing graded return to work for work hardening to avoid re-injury provides a record keeping penalty...not a reward.

In closing, there are many demographic factors that affect injury statistics, including work ethic, labor relations, medical services, intensity of letigency, and many other factors. Comparing operations in different locations, even if they are in the same business, is somewhere between tricky and impossible. The people in the trenches recognize the differences, and using safety records as drivers makes them cynical and often demoralizes them.

2. Negativity. This feature requires little explanation or justification. Behaviorists have reminded society of something that many members learned long ago but have lost. It takes about ten positive reinforcers to have the impact of one negative. Using negative reinforcement is powerful and shapes behaviors in a way unlikely to have the desired long term effect.

3. Superficiality. Every person who has been in safety work for any time at all is familiar with the accident pyramid. It doesn't matter how steep the sides are. If safety professionals would only work on fatalities and lost time cases and even recordables, they would have a relatively small sample to work with and it will often send them lurching in the wrong direction. Data to examine minor injuries and near misses is rarely available; and when it is, it is even more suspect than the information discussed before. If the data is complete and accurate, the people who matter probably won't look at it; and if they do, they won't understand it or draw the right conclusions and act on them.

4. Credibility of Investigations. In the hundreds of companies for which I have worked and the thousands of locations I have visited, I don't think I have ever seen consistently excellent accident investigations. If the LAPD can't do a credible investigation of the crime of the century, what chance is there that an overworked supervisor of fifty employees is going to conduct an insightful investigation of a case of tendonitis? The problem here is that the employees usually know more about what happened than what gets reported and often see the exercise as a sham. Reported causes are usually simplistic, if for no other reason than to comply with forms and electronic data processing requirements.

5. Inscrutability. Few really understand what accident statistics mean. Managers feign understanding so they won't appear ignorant. The higher in the organization, the less the understanding and the greater the subterfuge. Remember, many of these people got to where they are not because they are smart or good but because they know how to look smart and good. Inside, they know they don't know. If safety professionals have trouble measuring, even using universal terms that are at the heart of what interests people, like dollars, what hope is there for LWDCs or OSHA incidence rates? If the people who want to be inspired and driven are turned off, what kind of a driver are accident statistics?

6. Consequences. Behaviors have consequences, but accident rates almost never have consequences for employees. Hence, they have at best a passing or academic interest in them. They will not be motivated or certainly not inspired or driven by them.

Line managers may be measured in part by their safety performance in enlightened companies. Unfortunately, not many companies have clearly defined for the managers how they get the right rate and have demonstrated a bias toward measuring and rewarding the methods rather than the results. The effect is cynicism, confusion, occasional paranoia, and, in some cases, fear. These emotions can prompt short-term gains but yield to long-term hidden rot.

7. Inconsistencies. Much of this has been cleared up with more people going to universal measurement. But there are still residuals of old systems. The old ANSI interpretation of lost time still surfaces at times. Certain industrial groups have unique measurement nomenclature and jargon. Workers' compensation verbiage gets entangled with accident rate talk. This isn't a big problem but does blow smoke into an area that is screaming for clarification, not clouding.

8. Confusion. The human family keeps trying to simplify life by adding complexity. The best way to simplify is to eliminate. Safety statistics are no different. Everyone is trying to find new ways to display data and use it in more meaningful ways. You can even find applications of statistical process control to safety statistics. They are confusing and the whole exercise is probably based on some false premises. I'm not the brightest person in the world, but I've spent over thirty years in this business and someone is paying me to write a book. If I can't fathom some of the variant accident statistical analyses, are they likely to be effective motivation for an organization to work more safely?

COSTS AS A DRIVER

The so-called risk managers and their cohorts, insurance brokers, insurance regulators, and insurance companies, have confused this area so badly it probably won't ever be unraveled. However, this is going to be the most difficult traditional motivation for me to criticize because I have used it very effectively many times.

Again, I can best illustrate my feelings on costs as motivation with some personal experiences. In 1979 and 1980, I began to notice incidence rates climbing steeply in our shoe and children's apparel companies, particularly in their operations in two states, Maine and Rhode Island. I kept issuing warnings but little really happened except some shared concern.

I couldn't understand why no one really seemed to care. Then, one day between Christmas and New Year's Day in late 1982, the picture started clearing up for me. I was on vacation and my leader, Ted Mullins, called and started by telling me that our CEO, Ralph Ward, was on vacation too. Initially, I couldn't see exactly what that had to do with me until he elaborated. He said Ward was taking his vacation in his office with his suit on and that he would like me to get my suit on as soon as possible and take part of my vacation with Ralph in his office. He did not tell me what the subject was but left little doubt about its importance or who had suggested the parallel vacation plans.

The next hour or so as I dressed and headed to the office was one of the most intense periods of self-examination in my life. You see, I had a few opportunities to interact with Ralph Ward in small group settings since that day some five years earlier in Perth Amboy, but they were always planned; I knew the agenda ahead of time and they were non-threatening.

When I arrived, I went directly to his office. He greeted me and began to unfold a story that I generally knew but didn't interrupt until I was sure he was done. The deteriorating safety performance had come to his attention. Not in the form of the safety statistics that were featured in bar graphs in the monthly Safety Reporter he and every other significant member of company management received, but in the final budget for 1983. It called for a total workers' compensation cost of $13.5 million. Even in what was then a well over $1 billion/year company with net profits over $100 million, this was a number to be concerned with; particularly since it had been less than a quarter of that in 1981 and was projected by the actuaries to grow to $18.5 million in 1984.

Since insurance loss costs go directly to the bottom line, our profits would be adversely affected in 1983 by over 5 percent due to the workers' compensation hemorrhaging and soon would grow to over 10 percent.

Ralph told me that he had been asking his staff support people and company presidents what was happening and he wasn't getting pleasing or understandable answers. He had decided that for the foreseeable future (probably six months to a year) I would report directly to him. My job would be to devise a way out of our costly dilemma and brief him on about a weekly basis.

The company was an early victim of what soon became a national and even international epidemic of soft tissue injuries, with particular emphasis on the upper extremities as opposed to the old back problems. These industries (apparel and shoe) placed high demands on the hands, arms, and associated body parts. They were doing business in states, Maine and Rhode Island, that had very permissive or progressive (pick the most appropriate word) workers' compensation systems. In the areas with plants, the medical communities had reputations for being less than excellent. (I'm trying to be kind.) The legal communities in both states were very bright, aggressive, and highly motivated. They had participated in creating a very hospitable workers' compensation litigation environment for themselves. As Ted Mullins had suggested when I painted this picture for him, "You have the worst of all worlds: bad doctors and good lawyers."

The first thrusts of senior management had been to blast the laws in Maine and Rhode Island. Several months before, Ralph had criticized them in a speech before a Maine industrial development group at which the governor had been present. It was reported as far away as in the Sunday business supplement of the *Los Angeles Times* and was covered in the *Wall Street Journal*.

My early efforts were aimed at trying to get the focus off the laws, doctors, and lawyers. Realistically, management couldn't do much to change any of them, certainly not in the short term. They did have numerous areas of opportunity that they could control. The job design, health care support, incident investigation, rehabilitation and return to work programs, and claims handling and case management were as good as most people had but woefully inadequate for the onslaught they were experiencing.

An initial assessment effort involved conducting a detailed examination of over one hundred employees who had been categorized as permanently and totally disabled. In Maine, compensation rates are generous and last a lifetime. Many of our pieceworkers were at very high pay rates and were very young. The potential tab for just these cases was well into the tens of millions of dollars. We invested about $250,000 in extensive examinations that include participation by medical doctors, chiropractors, chemical dependency experts, and psychomotrists, and headed by a nationally recognized forensic psychologist, Phil Haber. Testing included IQ tests, interest tests, and use of the Minnesota Multiphasic Personality Inventory.

It was found that a few cases were exaggerated, more of them were understated, but most were as described. The etiology was quite complex and would require a very sophisticated response. Clearly the legal/medical environment had facilitated and precipitated the current problems. But, the company had sown the seeds of their own dilemma.

Ralph had another experience that moved him toward his eventual orientation. He was compulsively honest and believed that a person's word was his/her bond. The Maine governor was attending a governors' conference nearby and invited Ralph to meet with him privately and suggested that they might be able to resolve some of Chesebrough-Ponds' (which was, at the time, the second largest private employer in Maine) concerns about the workers' compensation environment in Maine. Ralph went; they talked; some conclusions were reached; the governor promised action (and confidentiality). A few days later, their meeting was widely reported in the Maine press with a perspective that was clearly that of the governor. In a meeting with Ken Lightcap, our Vice President for Public Affairs, Ralph was incredulous. He reiterated the agreement he had with the governor. Ken suggested to Ralph that the governor was a politician and undoubtedly played with a different rule book than Ralph did.

After several months of analysis and evaluation of options, I prepared to take the results to Ralph with my recommendations that would cost over $3 million. Some of that was a one time expense, but some of it would be ongoing over several years. At this point, I was working with a Vice Chairman, Dick Scheifele, who presided over all company manufacturing, engineering, and most technical staff services. He helped me get ready for the presentation to Ralph. Over lunch, the day before the presentation, Dick, always the insightful, practical executive, asked me what I was prepared to give up. He was assuming that Ralph would ask tough questions and attempt to pare down the initial proposal.

I told him that if Ralph started to do that, I was going to ask if we could return the next day. Dick looked a little puzzled and asked why I would do that. I said so I could pad the numbers to allow Ralph to go through the traditional cut-back exercise. Dick smiled at me and didn't say anything but I think I could read his mind. He was amused and was willing to let me place *my* career at risk.

At the meeting an interesting thing happened. And, I had sensed that it would. In the midst of a company hiring freeze and regular budget cutting exercises, Ralph approved all the expenditures, including numerous new jobs to address ergonomics, rehabilitation, case management, training, and claims handling. I sensed Ralph had come to a very rapid and accurate diagnosis of what was happening to us. He had not gotten to where he was without a great capacity for understanding when he had the facts.

In several preparatory meetings, Ralph had closed the meeting asking me, "What is the right thing to do?" He had moved past the laws and malpractice and even the costs.

He always knew that we had a responsibility to our employees and now knew that we had not done all we could to discharge that responsibility. He was prepared to make corrections and spend whatever we needed to in order to do it. He was no longer driven by the events or even the costs. He didn't want us hurting any of our employees if there was any way to avoid it. Most executives can probably be led to this point, particularly if they come to realize that it is also the best route to a healthy and profitable organization.

Within two years, our workers' compensation costs had dropped to below $3 million. We had world class ergonomics initiatives and return to work programs, as well as all the employee sensitive activities designed to prevent injuries by early intervention. *We had saved millions of dollars, but we were driven by a concern for our employees. I'm convinced that when we stopped being obsessed with the law and even the costs and focused on doing the right thing for our employees, we turned the corner.*

However, costs can be used as a powerful benefit in an organization constituted to pay stockholder dividends. This text has discussed how that works and it will be covered in some detail in Chapter 11, which focuses on the importance of selling benefits as a value. My organization rewarded me with the highest percent bonus any staff person had ever received. I was able to pass on numerous benefits to people who had worked with me. The primary medium of reward in corporate America is still money.

MANAGEMENT DIRECTIVES AS A DRIVER

We had used management directives before the story told above had unfolded. They had done little to change anything. They are primarily a source of amusement for employees. At times there are token reactions, but they don't last long and are purely cosmetic.

Another personal experience will illustrate why I don't think much of management directives. In our foods business, we did canning. It is a very unique business as it is seasonal and has its own culture. The management works unbelievably hard during the harvest season but is pretty laid back the rest of the year. They are inbred and very self- assured (and I might add competent and resourceful). When something needs to be done, they find a way to do it.

Our plant in Merced, California, had the worst safety record in the company. Being a firm believer in the axiom of greasing the squeaky wheel, I started spending a lot of time there. With an average of far less than two hundred man-years worked each year, they averaged over one hundred recordable injuries, or an incidence rate of well over one hundred. It was appalling and even as a large corporation with great buying power, we had to accept very unattractive terms from a local insurance company because none of our normal sources wanted anything to do with this plant.

General manager Ed Harmon, production manager Max Thomas, and I became well-acquainted. They seemed sincere in their desire to reform what they were doing and get a better result. We installed all the artifacts of traditional safety programs. Nothing happened. I was more perplexed than ever.

Suddenly, I noticed that for three months, recordable injuries dramatically decreased. I visited the plant with the express reason of researching what had happened. I met Ed and Max in the small lunchroom and couldn't wait to pop the question. I blurted out to Max that I was aware of their dramatic improvement and asked him to explain. He responded that he could but wouldn't.

Ed, his boss, looked at him, became a little stern, and reminded him that I had become a trusted friend and helpmate and that he had no reason to hide anything from me. He instructed him to respond to my question. Max got a little twinkle in his eye as if he had elicited exactly the response he was looking for and, as if on cue, looked right at Ed and responded that Ed was the explanation.

I became quizzical and provided the next expected cue for Max, asking him to explain. He related that in the past, about once a month at staff meetings, Ed went into a tirade about the bad safety performance at the plant and encouraged better performance or at times berated the managers and supervisors about malperformance in safety. He wrote directives, but not much really changed. When they met in the hall and out in the plant after the meetings, the managers and supervisors would laugh about what Ed might try next. The whole thing had just become a part of their culture. They not only accepted it but had come to like their poor safety performance almost like war veterans or athletes see a physical affliction, such as a gnarled finger or scar, as a badge of accomplishment.

Max then related what had changed. Suddenly, Ed stopped getting mad and issuing directives. He decided to get even...literally. He cut pay increases for several supervisors who had weak safety performance. He required accident reports to be on his desk within twenty-four hours, and more importantly, he started to read them. If he couldn't understand them and know what had happened and how it was to be corrected, he called the supervisor in and reviewed the incident and then visited the site with them, dropping everything until the matter was resolved to his satisfaction. Max talked about a lot of other changes in the management style of Ed Harmon as it related to accident prevention. They could be summed up by rejecting directives and attending to process, with a vengeance.

I was curious as to what had changed Ed. Research at corporate headquarters revealed that his boss, the head of manufacturing at Ragu, Gerry Chrusciel, had modeled the behavior for Ed that he expected of him. The very things that Ed was now doing had been practiced on him. Example is indeed a powerful teacher. Later in Chapter 14 on the importance of empowerment, I'll discuss my relationship with Gerry. The follow-up on this story is that the performance was sustained and a few years later, the plant won the coveted Chairman's Safety Award for the plant with the best safety performance in the corporation. I see the moral as: Directives don't work, *properly motivated* process change does. Ed was always sincerely concerned with his employees well-being, he just didn't know how to manage the change he wanted until he had a good example set for him.

INSPECTION AND AUDITS AS DRIVERS

I am somewhat ambivalent about inspections and audits. At their worst, inspections and audits are the worst of any driver. At their best, they are almost as good as the traditional drivers get and are useful in more progressive approaches.

At their worst, inspections only evaluate physical deficiencies and are used as a vehicle to leverage other agenda items and are rarely followed-up. Under this scenario, they would be better described as brakes rather than drivers for an effort to eliminate undesirable outcomes. They end up serving as proof that management doesn't really care about safety. Workers worst fears are realized. Hope is squashed. Deficiencies are detailed, yet nothing happens.

At their best, audits, evaluations, or assessments (depending upon what you choose to call them) can be effective if not essential management tools. They take the temperature of the organization to insure that all is well and that what is supposed to be happening is in fact occurring. One of the quality gurus has suggested that inspection is the Bat Masterson of the quality process. If properly conceived and executed, it can insure that the safety process is well-oiled and running smoothly.

Properly done, they must include behavioral and process examination. Well-thought-out protocols that have total organizational support are essential. Auditors must be well-qualified and trained. They must have the time and resources to do the job thoroughly, and the follow-up must be conducted at the highest levels of the organization, with deficiencies addressed forthrightly. That all seems pretty obvious but doesn't happen in many organizations. Chapter 18 on Measuring Results discusses more details of how this can be done best.

THE CORRECT DRIVER

The only really good reason to do safety is to prevent depreciating employees in any way (see page 98). This must become an organization value if resources are to be optimized. Management must believe that the greatest asset of the organization or extended family are the human resources and that successes are closely related to the caliber, health, and well-being of those human assets. They must protect them from any kind of loss or depreciation of their quality of life.

The Right Reason To Do Safety: Value Your Employees

- It Projects Sincerity: Employee loyalty is encouraged.
- It Promotes Teamwork: Trust develops and enriches teams.
- People Relate to It: Employees recognize leadership empathy for them.
- It Is Cost Effective: Employee contributions are optimized.

This is the right driver because:

It Projects Sincerity

Once employees come to think that the persons they work for have a real interest in them, they will start to trust the employers and then desire to work harder for them and follow their instructions.

It Promotes Teamwork

The current industrial buzzword is team. Teams don't work together very well if they don't trust each other. If self-interest is paramount, trust doesn't have much of a chance to prosper or even survive.

People Relate to It

Human beings relate to concern and empathy. They are repulsed by insensitivity and callousness, particularly by someone from whom they expect more.

It Is Cost Effective

It sometimes costs a little to communicate with employees and show them that they are valued. The return on investment is difficult to precisely measure, but those who don't pay now invariably pay later at a very high interest rate.

SUMMARY

The way the right driver is implemented is to make the people the first thing talked about at every safety meeting. Statements should be made verbally and in writing by authority figures and supported by every action. Chapter 17, "Installing the Process," discusses in some detail the techniques needed to show that management not only plans to do safety for the right reasons but is committed to it.

In summary, all the reasons to do safety that were criticized have some merit. They just aren't powerful enough to propel and sustain an optimal effort. Using them without the foundation of a sincere interest in the well-being of employees that is communicated to everyone, is hollow and ineffective. The reasons to place concern for employees first when it comes to achieving an accident-free working environment are few yet very powerful.

6

SEE IT AS PART OF THE WHOLE

"Tunnel vision is the degenerative disease of specialists."

Almost every recent talk or publication dealing with how to manage safety suggests that it should be integrated into how to manage the business overall. I've *never* seen anyone who does it *really well*. Why is that? With all the smart people who have tackled this, why should I succeed when they have failed? My hope lies in my vantage and the unique position VIRO could occupy in a company.

This chapter will start out by trying to figure out why integration hasn't worked. Then it will give solutions—how I think it can be fixed (see page 109). Finally, it will suggest why integration is essential to optimize the right outcomes, although for some readers it may seem like preaching to the choir.

Why Hasn't Safety Been Integrated into the Management Process?

- Safety is a "Johnny come lately" to most environments.
- The word "safety" is overused, abused, and misunderstood.
- Not many people get formal education in the subject; particularly those who need it the most. (e.g., business people).
- The breadth is not defined.
- The problem (losses) is poorly defined and generally misunderstood.
- Safety has attracted practitioners who are their own worst enemy.
- The obstacles encouraging segregation have proliferated faster than the impetus for integration.
- *No one* has really gotten serious about this.
- *Everyone* is scared.
- Change is hard; absolute change is absolutely hard.

Why hasn't safety been integrated into the management process?

Where Did Safety Emerge on the Time Line?

Have you ever seen one of those history of humankind time charts? Texas A & M houses one on the wall in one of their buildings. It covers the last six thousand years of human history. I was there to give a talk about fifteen years ago and when I saw it, it was all I could do to tear myself away and attend to my speaking assignment. If we superimposed safety as a distinct discipline, it would only go back about 1/100th of the way on the mural from right to left—only a few inches on a twenty-foot mural. For comparison, business, law, politics, and education were all highly developed disciplines by the golden age of ancient Greece. When someone discovers something (like safety for example), it is hard to blend it in. You have to figure out what it is before you can decide where to put it, right? We really don't know what safety is.

The Word "Safety" Is Not Clearly Defined

In Spanish, one word, *seguridad*, covers both personnel safety and security. Campus security officers are often referred to as public safety officers, although they really do security. Some police departments are in public safety departments. Safety is tacked onto the front of everything from pins to glasses, often producing half-truths. When people talk about "safety," it seems to mean different things to different people: a homemaker thinks one thing; a college student thinks another; a factory worker thinks yet something else. Actually, it may mean different things to factory workers in different industries.

Formal Safety Education Is Almost Non-Existent, Particularly Where It Is Most Needed

Until a couple of decades ago, few if any accredited college degree programs in safety existed. Even now, not many exist, and they are obscure, mostly in smaller schools and constantly fighting for their lives. Almost no curricula for doctors, lawyers, accountants, engineers, or business students include any instruction on safety, much less a full course. If they are told anything, it is mostly wrong minded, yet portrayed as authoritative—the worst of all worlds. There are very few programs for risk managers and almost no one knows what they do. Perhaps they don't even know. They mostly try to isolate themselves instead of integrating. Ditto for the fire service and fire protection engineers.

The Breadth of Safety Is Not Well-Defined

The end of the last paragraph leads right into this statement. What is included in safety? Does it include fire protection? If so, is that both automated systems and manual fire fighting? Where does industrial hygiene fit in the mix? How does that relate to occupational medicine and how do both together relate to safety? Back to the risk managers—what do they do anyway? And if they do more than buy insurance, what is it? Does claims handling belong with insurance, or is it loss prevention after the fact that better belongs in the prevention (safety) bag? Are wellness and employee assistance programs really loss prevention programs, and should they be managed by a loss prevention professional? How do environmental concerns factor into this mix—are they in the clean-up or prevention business or both...and who knows anyway? There are a lot more questions, but you probably get the point. These questions don't have answers that are generally accepted and practiced.

The Problem (Losses) Are Not Widely Understood, Categorized, or Quantified

Safety/loss prevention is generally misunderstood in kind, nature, and scope. The few decent publications that provide some insights concerning loss experience are ignored; discontinued due to lack of readership and reaction; or read by a very small segment of the population and then, more for academic interest than to formulate plans to do something constructive about the problem.

The *Cost•Of•Risk Survey* discussed in detail in Chapter 11 draws looks of consternation when mentioned to people. It may have been discontinued due to lack of interest and support. It is filled with useful information. The publication *Injury in America* was prepared by well-educated people (but, interestingly, no safety people), and it paints a picture that screams for action. Yet I've never met anyone who has read it. I'll reference it later. Health offenders (i.e., smoking, alcohol, obesity, etc.) result in loss but are rarely considered by loss prevention experts. Doctors and politicians deal with these, but safety professionals could surely improve their effectiveness. The costs keep escalating and good ideas are still few and far between. Confusion and consternation are more prevalent than edification and action.

Safety Professionals Are Their Own Worst Enemy

As a category, they are among the most family-oriented, clean-cut, wholesome-living, nice people in the country—actually in the world, based on my experience with them in many other countries. I am constantly impressed with their goodness. That's the good news.

The bad news is that they are very bad at telling their story. They tend to enjoy each other and frequently talk to each other but not to the outside world and often not even to those they work with. And when they do, they are so caught up in helping people that they rarely see the big picture and have difficulty relating to the concerns of the rest of the business community or the world.

Safety professionals normally are not interested in stature. If you don't seek it, people rarely offer it. Hence, they don't get it and end up without much influence. They need an advocate.

Business Keeps Moving Toward Segregation Instead of Assimilation

As fields fragment and technology in areas such as occupational medicine, industrial hygiene, ergonomics, and many others deepen, integration into the main stream of life, politics, and business looks further away rather than closer. Since this is an interdisciplinary, multi-functional activity that requires focus, safety managers may be headed in the wrong direction. They are diffusing and confusing rather than clearing up and combining. Even within their own profession, they tend to be fragmented. They are fragmented by industry and discipline, which often limits impact. Even the National Safety Council encourages a form of segregation with sections by industry. Every discipline has their own society, and getting them together is not easy. They have tended to grow apart instead of together. Very few people move from this business back into the main stream. It is so cozy and enjoyable, they just stay. Isolation is perpetuated. They need a common ground and language...and purpose— directed by a unifying process.

Is Anyone Really Serious About This?

The rhetoric is profuse. The useful, directed action is fragmentary and often at crossed purposes. Who is the champion? Who is the authority? It isn't the insurance industry. There are many reasons why a book should be written about it...but someone has already done it. His name is Andrew Tobias and his work will be discussed later. It isn't government; they are so confused, bureaucratized and politicized that the fog will never clear. Our best hope is, and probably always has been, industry. But they are generally consumed with other things, like making money in a very competitive environment with government and labor (unlike our competitors in the rest of the world) throwing up road blocks at every opportunity, in spite of claims to the contrary.

Fear and Confusion Predominate

Different groups are afraid of losing influence. Safety professionals are afraid that they may be a dying breed. Youths are afraid of nuclear war—they may be more insightful than everyone else. The Russians may be confused, but they are not defanged; international terrorism is rampant. Terrorists are learning about nuclear energy, and nuclear terrorism is probably just around the corner. *These are serious hazards and safety problems.* Parents are afraid of hamburgers and the asbestos in their children's schools. They are also taking courses and giving their children instructions on how to deal with "strangers." They are learning family code words. Men and women who are good at what they do and who show up at work every day are afraid of losing their jobs as professions become obsolete or non-viable. All minorities are afraid of all majorities, and nobody really knows who is which. Some majorities are afraid of some minorities. This book's intention is not to minimize any of these fears. This is a scary world. Most people don't perform well (or at least not logically) when scared. The very nature of the safety profession is to alleviate fear, so if they can overcome fear themselves, they can get on with the business of curing others' fears.

Change Is Hard; Absolute Change Is Absolutely Hard

This is a play on the saying that "Power corrupts and absolute power corrupts absolutely." Let me tell you about my shower head. I am 6' 2" tall; my wife, Barbara, is 5' 2" tall. This causes problems with cars and shower heads. We have our own cars and now they have programmable seats, so we're in good shape there. No such luck with shower heads. I'm supposed to turn the head aside so when she turns it on, it doesn't shoot out the door at her. I never did it. She threatened to take numerous severe sanctions. I got scared. We negotiated and compromised. She prepared a calendar/chart with stars for me to affix each day when I got out of the shower and turned the head. I think I got twenty-nine out of thirty-one and we were both pleased. The chart came down and I reverted back to my old behavior. I really want to do the right thing. My brain just isn't programmed correctly. She understands that now and the sanctions are history. I'm trying harder, but my success rate is still not up to my won-lost percentage when I pitched college baseball. We laugh about it now, and, occasionally she still gets wet prematurely.

Seeing safety in its broadest terms and getting safety practitioners to accommodate that thinking and prepare to respond to it won't be easy. Until this is done, the opportunity for integration, much less synergy, is far beyond reach. Slow change hasn't worked in the past. There is no reason to believe it will in the future. The industry may need a quiet revolution.

SOLUTIONS

Some solutions are described on page 109. These are what are needed to integrate safety. Some of them are tongue in cheek.

This whole VIRO process calls for a lot of participation on your part. If you decide to try VIRO and embrace this value, sort through these solutions and try the ones you like; reject the others.

Integrating Safety/Making It a Part of the Whole

- Obsolete the safety engineer as we know them.
- Obsolete the safety practitioner completely ...eventually.
- Start writing interesting books about this subject.
- Get someone important interested in loss prevention.
- Institute a national education program.
- Create good reports that we can all understand.
- Inspire courageous safety professionals.
- Catch a vision of how important all this is.
- Develop a strategy to stop unnecessary losses.
- Find tools and tactics to implement the strategy.
- Come up with a plan...then sell it and integrate it.

The Safety Engineer Has Been Obsolete for Years

Now the industry needs to recognize it. Safety engineering isn't like the use of electricity or the manufacture of chemicals. It is not an engineering activity. Devising methods to protect physical and human resources from hazards is a many faceted endeavor that requires, among many other skills, some occasional engineering aptitude. The work is seriously depreciated if engineering is emphasized. Safety "engineering" today (and in the future) is about thinking, creating, and communication. The technical trinkets are the easy part.

Plan to Eliminate the Safety Practitioner

Those of you who are for term limits don't need to read any further on this one. *All* safety jobs will *never* be eliminated. But the objective should be to teach people how to do it; then get out of the way and let them do it. Safety is not an end in itself. In almost all venues, it is a means to an end. The most enlightened practitioners, and the empowerment directed workers and managers they serve, see this clearly. The self-directed safety empowered organization of the future will have very little need for a safety practitioner. Everyone will be a "safety advocate" and all employees will have assimilated the skills and know-how. Safety will be part of their make-up.

Write Some Really Good Books About This Subject

Those books should tell people how to do it themselves and then offer the needed tools. They should be easy and yes...even fun to read. Otherwise, they won't get read. When is the last time you finished a book that you couldn't understand or that you found profoundly dull? Even if you have, you probably can't remember what it was about and surely never acted on it. I'm taking my own advice on this one. I only hope and pray that the work of a lifetime and the encouragement of wonderful friends and family can overcome the lack of literacy of an indolent chemical/safety engineer (ouch). You judge. *Write me and tell me the verdict.*

Get Someone Important Interested in Loss Prevention

If safety professionals could get Newt going on this one, they might have a chance. Furthermore, it might help him on some of the other problems that he has a good start on but may have missed some key points. Seriously, until safety gets a captain with real horsepower, serious, broad-based loss prevention is dead in the water. Previous chapters discussed Chesebrough-Pond's CEO, Ralph E. Ward. When he caught safety, he spread it throughout the organization. This book will explain how later. Safety needs advocates with powerful spark, spark that can get VIRO and other loss prevention techniques spreading like wildfire.

Institute a National Education Program

The areas in which our society has achieved spotty success on loss prevention are those in which we have triggered off a national education program. Real progress has been made on curbing drunk driving, children's head injuries while bike riding, car injuries related to lack of seat belt use, illnesses related to smoking, and premature death related to obesity. Sure there are laws covering some of these areas, but what prompted the laws? You guessed it, education. Education is part of the answer to all of these problems. In response to the question, "What is the good life?" Plato develops the thesis that the life of reason is the happiest and best. He bases his argument on the Socratic dictum, "knowledge is virtue."

With loss prevention, safety professionals need to "act locally but think globally." They say very little on the process or mind-set needed to minimize unnecessary loss although they make a lot of national noise. Some people still actually think some others are "accident prone," as if the problem is genetic. In reality, folks who have a lot of accidents are "accidents waiting to happen." They have a life style that sets them up for unnecessary loss.

Create Good Reports that Everyone Can Understand

When is the last time you saw an accident or loss report that meant anything to you? We need to put Ross Perot on this one. Look what he did on the deficit. No one knew what it was or what it meant in early 1992. Several Larry King Live shows and numerous charts and bar graphs later, everyone had *discovered* that it was our nation's number one problem, even Bill Clinton and George Bush.

Unearth, Train, and Inspire, or Resurrect Bright and Courageous Safety Professionals

There is a fine line between courage and stupidity. Almost every time someone has suggested that I have been courageous, self-examination has forced me to admit to myself that the real cause for my apparent boldness was a profound lack of knowledge of the possible consequences of my behavior. That is not always bad. Boldness will be needed to birth VIRO. In spite of the receptive atmosphere described in Chapter 2, it will take a unique kind of safety professional, one well-placed in a first class organization to get this going.

Catch a Vision of How Important All This Is

Without a vision, nothing happens. Speech/language and abstract thinking differentiate humans from their relatives in the animal kingdom. People now know animals can do work but can't contemplate their own destiny nor do they have a vision of the future. The creators reserved those things for us. These unique endowments are of little use to us if we don't apply them. *Remember, if you don't use it, you lose it.* This subject (i.e., having a vision) is a sleeping giant. Let's wake it up. VIRO could provide the wake-up call.

Develop a Strategy to Stop Unnecessary Losses

Many do this individually, without a plan. They need one. They could use a "safety czar," a leader like former Surgeon General C. Everett Koop. Look what he did for smoking cessation as a one man gang. Just think what he could do if set loose on this bigger subject with a good support system. We can't just hope this will happen; we need a *national* strategy.

Create Tools and Employ Tactics to Implement the Strategy

A strategy is just the first step. Then tools are needed to implement it. I hope VIRO will be one of them. Once we have the tools, the tactics will become more apparent.

Come Up with a Plan...Then Sell It

Here's where selling skills become critical. Don't look at me. I'm great at buying things. My basement and garage prove that. I just can't sell anything. Traditionally, safety professionals all have this problem. We need help or to identify someone among our number who has latent talent.

WHY WE NEED TO INTEGRATE SAFETY/SEE IT AS PART OF THE WHOLE

Consider the alternative to integrating safety: business as usual. How are managers doing? They've hit a wall. They need some better ideas. They need to work smarter instead of just working harder. They probably can't work much harder anyway. There is an interesting parallel. Often the most dedicated employees, the most resourceful workers are the ones who get hurt. Is that because they didn't work hard? Of course it isn't. It's because their values are slightly out of whack. The only way to fix this is by building a broad base. If safety is isolated, that will never happen. When someone is put in charge of something, everyone else assumes it's not their job.

The clues that integration is beginning to happen are shown on page 115.

Clues that You Are Achieving Integration of Safety into the Management Process

- There are very few identified safety practitioners in or near your organization.
- Very few people know the safety practitioner if you have one.
- There are no safety contests...and nobody wants one.
- There are no safety committees...and no one sees the need for one.
- Responsibilities for specific safety subjects like OSHA compliance, ergonomics, industrial hygiene, fire protection, claims handling, environmental compliance, have been assigned to specific line managers who have developed expertise in the area and know where to find help if they need it.
- Loss prevention is a part of every discussion. It isn't always covered first but it is never overlooked.
- You know what your safety values are and you know how to measure them and correlate them with your performance.

I've always liked clues. Sherlock Holmes used clues to solve crimes. Our crime is unnecessary losses. How do you stack up against the clues? Do you have a chance to solve the crime?

By the way, these clues are a package. If you are okay on the first four but not on the last three, you may be a worst case and beyond help. Check your pulse.

7

RECOGNIZE THERE IS NO END

"Eternity has no gray hairs."

The quality control advocates have sold the idea of continuous improvement, although it has not been internalized in many places. The quality philosophy is that efforts cannot end. Likewise with safety efforts...there is no end. For loss prevention advocates to take any other position is self-deluding and guaranteed to produce nothing better than the accident cycle. Let's take a look at this from philosophical, theological, and practical standpoints.

PHILOSOPHY

Only philosophies that take a long view are ultimately useful. The popular ideal of live for today and tomorrow will take care of itself, is pure Sophistry.

Personally and industrially, Americans have become selfish and short-sighted, with over-emphasis on the next quarter's

PEANUTS   By Charles M. Schulz

PEANUTS reprinted by permission of United Feature Syndicate, Inc.

profits. The problem with short-sightedness is illustrated here with political and business anecdotes. People must turn away from industrial myopia and the selfish culture underpinning it. Managing safety through a foundation of values as the leading edge can prompt refocusing.

At the risk of dating this book, we'll take a look at a perfect example of near-sightedness in the political arena. The "non-essential" government was shut down due to philosophical conflict between President Clinton and Congress. Most people didn't notice. Allegedly, the issue is a balanced budget. In reality, the argument is whether older citizens pay as they go or leave a bill to be paid by their children and grandchildren through the national debt. Our legislators have been taking the short view and growing our national debt.

The real issue has been masked and misstated. An appeal to the humanity of the older generation would surely be very favorably received. When placed in correct terms, most people probably don't want to leave mortality knowing that they have bequeathed a burden for their children and grandchildren. Their basic instincts are just the opposite. Look at the trust funds and other legacies that most people of all cultures and means leave their children. They usually only take the expedient course when they don't fully appreciate the implications.

Business usually lacks familial ties. More often than not, executives do not have close personal ties to their stockholders or employees. So the long-term outlook is easy to ignore. They need to look good *today*, even if it means leaving a mess for those who follow. They will be long gone with their stock options, retirement benefits, 401(k) plans, and perhaps even "golden parachutes."

Come on, is that all we can expect from the leaders of American industry? I knew some that didn't behave that way. You don't usually get more from people than you expect of them.

Let's start exerting influence and telling our leaders that we expect more, through phone calls, participation at stockholder meetings, letters, and so on. We want them to act responsibly and consider not only our future but that of our offspring.

THEOLOGY

All members of the human family have a deep spiritual base. A few try to deny it from time to time. Some longer than others. But virtually everyone has at least one "foxhole conversion" in their lives. To look at any subject that has a predominant human element in it and to not consider the theological reference points is to be like an ostrich.

Almost all spiritual thinking includes contemplation of what happens after death, and many religions have formal theology defining afterlife. Accountability for mortal activities often plays into post-mortal status. Sometimes the vision and linkage is clear and well-defined; sometimes it is vague. But accountability almost always looms large in formal belief systems and, more significantly, in the minds and hearts of the believers. Surely it affects their behaviors. So theology stresses the importance of the longest term (eternal) effects of behaviors.

Most people can easily transfer these deep understandings of long term consequences to safety or any other subject they may consider.

PRACTICALITY

Philosophy and religion are not much good if they cannot stand the test of practical application. The best philosophies and religions prove themselves true through their utility in everyday lives.

My first job with the Factory Mutual Engineering Division started with an interview with the District Manager, Al Booker. Following the meeting, I resolved to behave as if I'd retire from FM in forty plus years. Two of my colleagues mused about "giving it a try." They didn't even finish the training process. It was almost four years later that I left, on the best of terms, for a wonderful opportunity with Merck & Co. I wasn't any more able than those colleagues, but I took the long view; they were operating day to day. The fruits of expediency invariably have a very bitter aftertaste and often produce indigestion. The conventional wisdom that pleasure and happiness are synonymous is hogwash.

I have always felt that the best investment I could make was in the education of my children. Now many people forego having children because they are "expensive, time consuming, and cause you a lot of trouble." I wonder how those people will feel when nobody is around to visit them in the nursing home. Or even less abstractly, to appreciate the fruits of their labors with them; not to mention missing the basic training parenthood provides.

COMMON PRACTICES THAT ENCOURAGE EXPEDIENCY

There are several common practices that encourage expediency. They are described on page 122.

Common Practices that Encourage Expediency

- Accident Reduction Goals—Focus is on numbers, not process.
- Cost Reduction Goals—Focus is on money, not people.
- Inspection for Compliance—Focus is on reports, not exposures.
- Incentive Programs—Focus is on selfish, not selfless.

Accident Reduction Goals

Accident reduction goals are usually short term and encourage "cooking the books" if the results don't seem to be predicting the desired outcomes. The emphasis is off process and clearly on expediency. Companies bring people back to work prematurely. They have a dim view of the manager's sincere concern for them. The companies don't provide work hardening or alternative work lest they have to count the case as OSHA lost time. So, they reinjure the person. Then they have to cover that up. It is like lying, and perhaps in the most perverse way: lying to themselves.

Cost Reduction Goals

When companies focus on workers' compensation cost reduction, the employees who are system abusers will be one step ahead of the employers. The sincere, hard worker will often not get an attorney, and can be manipulated. If a few dollars are saved on a settlement or some medical benefits denied, companies enhance the chance of meeting their goal. When the word gets out, more people get lawyers. The "comp" culture spreads. Pretty soon there is an alienated group of employees who are cynical and even more resistant to long term cost control. But, they met the goal for the year.

Inspections for Compliance

Much of so-called safety inspections are essentially "housekeeping" surveys. At best, they end up spending their time on the symptoms. At worst, they waste time on things that have very little to do with conserving physical and human resources. Sure, tidiness and organization correlate with loss prevention. But how often is it the proximate cause? Will the recommendations just fix it now or do they include process changes that will correct the problem forever?

Look at training programs for compliance. Do they measure who was *in* the training class? Or, do they critically determine what needs to be known to avoid loss; then teach the skills needed; then have the skills demonstrated and coach the students; and finally, insure the objective has been attained by monitoring and evaluating performance? That *is* harder. But it produces the right long term results and sends the message, we know and we care; we expect excellence and we're willing to pay for it. Very few compliance inspections will go beyond checking who signed the register or who sat in the room during the session.

Incentive Programs

Everyone knows that compensation (pay) is a satisfier, not a motivation. But they can't seem to see incentive programs for what they are. They encourage short range performance followed by periods of complacency...unless you keep raising the ante. How high can you go? How great is the disappointment when it all, inevitably, comes to an end or when there is some disagreement on what constitutes conformance to the goal to get the reward. *All* incentive programs need to be *judiciously* stopped. Soon. These are just pabulum for the immature. What we are talking about here is a *quest* for *maturity.*

LESSONS FROM THE QUALITY MOVEMENT

U.S. business took almost fifty years to recognize Edwards Deming's good ideas. I love the story my colleague, Barry Jessee, tells at our TQM seminars for safety professionals.

He tells of a large U.S. computer company who specializes in mainframes. They order a part from a Japanese company and specify that they want less than 0.001% of the parts to have defects. They order ten million parts. Numerous boxes arrive with the parts and one is distinctively marked. The person receiving the parts sends that one directly to the Vice President of Purchasing. It has an important looking note on it from the president of the supplier. The note apologizes for any possible misunderstanding and expresses confusion about the purchasing specifications. Then it indicates that this particular box contains the one hundred defective parts requested in the purchase requisition.

This would be funny if it wasn't tragic. We just don't seem to get it. It's about values, stupid. The Japanese were converted to doing it right the first time over fifty years ago. We are going through the conversion process. If you have any doubts about that, just drive and closely examine comparable U.S. and Japanese auto products. We're gaining but there is still a gap. The answers are in what we believe in; how that translates to our national values about making good products; how that transforms our national culture; then what we do about it. It is mostly management's fault because they run the place. But the rest of us need to get on board.

WHAT DO WE DO?

We must vigorously encourage the long and broad view of safety. The following pages provide a checklist with which to start.

Overcoming Immature Focus on Expedience

- Focus on the trip instead of the arrival.
- Put *pressure* on process, not results.
- See every unwanted result or bad outcome as being the result of an *accident*.
- Develop an organization strategy for loss prevention that has no end.
- Develop and communicate a vision of twenty-first century safety.
- Deliver on the vision.

Focus on the Trip Instead of the Arrival

The statistics are the arrival. The *trip* is what produces the results that become numbers. After shredding all the accident statistics (except the ones OSHA requires and one set for top management) start measuring behaviors, attitudes, values, and other *predictors* that people know correlate with reduced accidents. Tell *everyone* about how you're doing against those criteria. Provide lots of "atta-boys" when you make progress. Rejoice together. This presupposes lots of involvement in determining what those behaviors, attitudes, and values are. A mundane example would be wearing hearing protectors in high noise areas. More subtle would be success in teaching concepts critical to avoiding wrong outcomes and the most advanced criteria—what would represent post graduate school in this realm—comprehensive measuring of safety values and responses.

Put *Pressure* on Process, Not Results

Once you have focused on the trip, you need to enhance and refine transportation and navigation. Do everything you can to get everyone understanding the *process concept* and the elements of a process that will predict the outcomes you want. Let *appropriate* peer pressure run rampant. If this is such a good idea, the people who matter most will recognize it. When it starts to work, they'll embrace it.

See All Wrong Outcomes as Resulting from Accidents

Broaden your view of loss prevention. Stop defining injuries as the only unwanted result of accidents. Any bad outcome is the result of a process failure, an accident. Your thinking should focus on the relationship between beliefs and outcomes and the process that connects them. You'll know you're there when you find yourself recording and being concerned with the number of process failures instead of the number of injuries.

Develop a Safety Strategy that Has No End

Write a new strategy. After you do it, red-line all items that have an end point. Sometime later, short term projects that can have end points will be discussed. But for now—until you understand the importance of this—avoid short term strategies.

Develop and Communicate a Vision of Twenty-First Century Safety

Every employee should start to feel what it will be like to work in an organization that really knows their employees are their bread and butter. They need to feel that they will be able to control their own safety and that management will accept the blame for process failures. This will be an organization that is a step beyond analyzing near-misses (better described by a colleague as "near hits") and a step beyond applying Frank Bird's (the father of damage control thinking) theories. Everyone will be secure knowing that they have a clear and concise understanding of what could hurt them and how to avoid being victimized. They know everyone else knows too. Certainty is a constant companion. Uncertainty is something they can't remember and is only of concern to those still practicing twentieth century safety.

Deliver on the Vision

Have you ever seen a sign that says, "Don't Even THINK of Parking Here!"? It is much more effective than any other I have seen. When I've seen it, I think, I don't know what they have in mind doing to my car if I park there, but I have visions of it in a crusher rather than just being towed away. I've never even *thought* of parking in a place with a sign like that. My advice to you is, don't even *think* of getting started on Recognizing There Is No End unless you are prepared to deliver. Employees are waiting for a management vision of safety that has no end. If you try and fail, they'll be *terribly* disappointed. They will have lost any confidence that they may have had in you.

I see this value, Recognize There Is No End, as *the flagship value* of my VIRO program. As time goes on and this method is applied to other broader subjects, like optimizing business (e.g., human resources and purchasing) or personal lives, this may end up being a flagship value for all of them. Only time and experience will tell.

8

FIRST, IT IS A PEOPLE BUSINESS; THINGS ARE A DISTANT SECOND

"As people think, so they are."

. . . .

"People are more easily lead than driven."

Early in my career, I really didn't know exactly what I was looking for when I went into a facility. So, I relied heavily on checklists and what my leaders told me to do. As I acquired experience, I began to modify my approach slightly to allow for more creativity in my evaluation and recommendations.

As I became seasoned, I almost felt that I could guess the incident rate or workers' compensation modification factor as I was walking up the path to the front door, or at least once I got into the lobby. That may be exaggerated, but I certainly had a good fix on those measures after simply talking to the person who managed the facility for a few minutes. The walk around, interviews, and record checking only confirmed what I already knew.

I am not gifted or uniquely qualified to perform program evaluations. Anyone with varied experience in this business for ten plus years who is sensitive and has drawn correlations in their minds could do the very same thing. I'm sure many of them have, perhaps better than I. What, then, does experience teach? That the facility manager determines the environment. What do you get out of an interview with a facility manager? You find out what they believe. You find out what they value. Just by observing their desk arrangement and what they have hanging on the office wall, you get clues. If you know how values translate into safe or unsafe performance, you've got it. Is this unique to the safety profession? Probably not. But in most cases, people don't stop to analyze how or why they are doing what they are doing.

Let's look at how a doctor diagnoses an illness. Specifically, let's look at asbestosis. At first I thought the person reading the chest x-ray could see findings equivalent to a sign that lit up and said "I've got asbestosis." Questioning radiologists revealed that any fibrotic condition of the lung looked the same on a routine x-ray, whether the etiology was coal dust, heavy smoking, silica, or other offenders. They were able to make only a differential diagnosis, or listing of several possibilities. By interviewing the person and determining work and exposure history, the possibilities can be prioritized. In fact, a careful interview of the patient is by far the most important part of making a diagnosis, whether for lung or any other illness or injury. All soft tissue injury diagnoses require not only physical examination but also knowledge of the job description and careful job observation. Often injury intractability can be directly traced to failure to do just that. My physician sons-in-law tell me that studies have shown doctors can be 85 percent sure of their diagnosis simply by talking to their patients. Physical exam adds another ten percent of surety. Tests and high-tech studies are only meant to confirm or aid treatment planning. What's my point? That safety (the business of ensuring wellness) is like medicine in that people are what matter—their stories, their behaviors, their symptoms, not machines/equipment/ technology operated by individuals.

In a recent conversation I had with a close relative, Bill Martin, we discussed how the VIRO theory applies in everyday situations, particularly in the prevention equation. He is a lineman for the local utility and is an accomplished ski instructor. He has many other interests including flying (he has a pilot's license), emergency health care (he is an Emergency Medical Technician), and property management. I asked him if he could tell which people were most likely to incur skiing injuries. He said, without hesitation, "Absolutely." He then proceeded to detail the behavioral characteristics and habits that predicted injury. He said all the ski instructors could spot those who were high risk for injury.

He didn't talk about weather or ski slope conditions although I'm sure they are factors. He dwelt on attitudes and behaviors of the people heading to the slopes as the major predictors, all linked to their values and beliefs. How can we use the connection to prevent skiing injuries? Unfortunately, not much at a public skiing area except to prepare to pick up the pieces. If we had a private skiing club, though, that had to buy liability insurance, Bill's know-how could be very valuable...if applied.

In his job maintaining and repairing power lines, Bill has observed his co-workers behave dishonestly or unsafely. We have had lengthy discussions about my VIRO ideas, and he verifies that his co-workers *behaviors* correlate well with their beliefs and with management values, as imparted to them by message, behavior, or system.

PARTIAL FAILURES-WHAT CAN WE LEARN?

The following details what we can learn from partial failures.

What Can We Learn from Partial Failures?

- Causation Theory—Oversimplification is misleading.
- Behavior-based Safety—We must go further.
- Empowerment Programs—They must be done correctly.

Causation Theory

Ever since the concept that eighty percent of accidents are due to unsafe acts was advanced, safety practitioners have recognized the need to focus on behavior instead of so-called unsafe conditions. This *apparent* understanding of what is at the roots of losses would seem to be good and useful. Unfortunately, it has created as many problems as solutions. Why? Because behaviors are only manifestations of the true roots-beliefs and values.

This misdirection began with the erroneous assumption that behaviors were the principal cause of losses. Any analysis that begins with an invalid assumption, no matter what the subject, jeopardizes the likelihood of finding sound solutions. (What do you think geometry proofs would look like if begun with incorrect assumptions? There would be a lot of bridges falling down.)

Another wrong assumption in loss prevention has been that there is a single proximate cause for every or at least for most single accidents. That is patently wrong, as anyone who has investigated a large number of accidents is well aware. Most safety professionals, including myself for several years, proceed on this incorrect basis. When I designed the first accident reports for Chesebrough-Ponds, guess what I did? That's right, I put two boxes on the report and, by report design, forced everyone who investigated an accident to mark the cause as either an unsafe act or an unsafe condition. When I think of it now, I am ashamed of myself. I had been in the business for over ten years. I had become wed to a wrong idea. The right idea is that almost every accident is the result of multiple causation—even the simplest mishaps. The major incidents not only have multiple causes but also several dimensions to each of the causes.

Real cause inquiry should ask, "Was it the job design; the tool design; the way you were told to do the job; the way you decided to do it; your failure to report your discomfort or pain; or the supervisor failing to respond?" This list could go on almost forever. Obviously, the answer is, "All of the above and then some." Unfortunately, if you find who did an unsafe act, blame them, and then trash them, you *think* you've fixed the problem.

Timing is another issue. Most accident reports also require giving a specific time when the event happened. Yet most industrial soft tissue injuries represent some form of a cumulative trauma. When did it happen? Was it the first time you did something the wrong way; the first time you experienced discomfort; the first time you felt pain; the first time the pain was so severe you told someone about it; or when you finally couldn't work any more?

How can accidents be reduced when everyone is unrealistic about what caused them? What can be learned from this obvious failure? Managers need to step back and take a whole new look at how accidents are investigated and get off the kick of labeling the cause either an unsafe act or unsafe condition. Until that is done, companies will be mired in confusion and condemn themselves to *running* down blind alleys. You know, they usually do *run*. Most accident investigations have an urgency about them that in itself often predicts a process failure.

Behavior-Based Safety

I've talked about the specific shortcomings of this method of accident prevention in Chapter 3 and almost painted myself into a corner. Now I need to work my way out of it. That chapter discounts behavior-based safety as the answer to loss prevention. Now this section will say that one of the values that will predict good outcomes is recognizing that a focus on people is essential to minimizing loss. The deconstructionists would have a party with what has been done and said up to this point on this subject, so let me explain.

Yes, people are the main ingredient in accidents. No, behavior-based safety is not the perfect answer. Let's look at the area where most fatalities in this country have incurred and probe the causes. Americans mostly die on highways. Each year we lose roughly the same number as were lost in the Vietnam War. If a wall were erected with all of the names of traffic fatalities each year, just think of the number of walls there would be. Picture it.

Now, why are they killed? Is it bad highway design? Is it bad braking systems or bad tires or horns on cars and trucks? Occasionally it is...but not very often. Two of the main causes are substance abuse and sleep deprivation. Are the causes people or inanimate objects? Let's apply behavior-based safety. The critical behaviors that cause the accidents are driving on the wrong side of the road, driving too fast, cutting in and out of lanes, backing up without looking, etc. Then people are told not to do those any more and recordkeepers keep track of how well they are doing, providing soon, certain and positive reinforcement when they drive on the proper side of the road, maintain appropriate speeds, get out and look before backing up, and only pass when necessary, and then at the right time and in the correct fashion. Bingo! The problem is solved and smaller walls with fewer names on them can be built. Does anyone think that will work?

Well, it certainly wouldn't hurt except to the extent that it obscured the real solution. The real solution is not as complicated. Just stop people from driving when they are sleep impaired or substance abusers. Can it be done? Of course. We just need to want to.

Let's look at sleep deprivation. One of the main culprits of accidents is truck drivers who are forced to drive long hours to meet on-time requirements set by people in suits in offices, most of whom have never driven a truck for twenty hours in a snow storm.

What would happen if everyone believed that a life lost in a highway accident had the same terrible effect on families as a life lost in a war…even one fought for a good cause? What would happen if they believed that they could prevent most of those accidents? What would happen if someone came up with a system to substantially reduce the number of people driving "under the influence" and those truck drivers driving sleep deprived? Suppose our whole society just said, "Enough, we don't want a big new wall each year." It could be done because beliefs changed, people started to have a new sets of values, and our culture changed. That resulted in different attitudes and behaviors that reduced the loss of life on the highway. See, behaviors are important, but as a by-product of changing beliefs, not because behaviors were directly manipulated. Behavior modification alone will work sometimes, for a little while. But it is *not* the solution.

Perhaps improved highway safety could be initiated by building walls and showing visitors crying and laying flowers. It has prevented the U.S. from getting into more Vietnam Wars. Maybe it would work for vehicular accidents.

Empowerment and Team Building Programs

The classic empowerment program is self-directed work groups. However, employee involvement is the watchword everywhere in industry now. It is also hot in other environments, as is "team." Will getting people involved reduce accidents? Of course it will. Is it the final solution? Of course not. Getting the wrong people involved at the wrong time in the wrong way won't optimize output. People must want to be involved or they are the wrong people. They will undermine the process and prove that you are wrong. You know that they can do that. How many times have you forced your children to do something and when it turned out to be a disaster, you've sworn never to let them touch your car, lawn, crystal, clothing, etc. ever again?

The wrong time is any time potential team members don't want to do it. Correct timing is as simple as that. Participants must be allowed to set the time. If you dictate time, it will usually be the wrong time. What is the wrong way? Any way that is less than optimal. Have you ever watched a 100 pound slightly built woman move a full fifty-five gallon drum, having done it for years? Give that same task to a 6' 4", 225 pound linebacker and he will probably struggle with it unless he has done it before. Why is that? Technique is the answer, not strength. People must understand how important method is to both efficiency and safety and quest after learning how to do it the correct way. Just leaving them alone or assigning them to a team they don't want, letting them figure out practices and procedures for themselves is not the answer. They must want to perform. (That implies that the task is at a time they find acceptable). And, they need to understand the value of using the best technique and be motivated to learn and apply the skill.

In 1995, Buck Showalter, manager of The New York Yankees, allowed Bernie Williams, his center fielder, to return to Puerto Rico to see his new baby right in the middle of the pennant race when Bernie was the New York Yankees hottest hitter. If Bernie believed that it was his duty to be with his wife and baby, how well would he have played that day anyway? Showalter did the right thing. In the long haul, the team was better off.

So what's the point? It is that self-directed work groups and other team processes are delicate and finicky. Are empowerment and teams part of the answer to bad outcomes? Sure, they are *part* of the answer.

SUMMARY OF LESSONS FROM PARTIAL FAILURES

What can be learned from pondering the Causation Theory, behavior-based safety, empowerment, and other "programs" that seem to focus on the human side of safety? Here are some ideas:

Half Truths Can Be Deadly

People and their problems are complex, and complex situations defy one line solutions. Safety professionals must take a broad view of the subject and base their conclusions and actions on a foundation of correct principles, beliefs, and values.

Simplistic Solutions Are Usually Littered with Pitfalls

This text highlights a few as they apply to areas to which we can all relate. Many of the readers could probably add to my examples. Feel free to do so. It will add some dimensions to your understanding of the concepts that are being exposed here.

Improper Motives Can Yield a Bitter Harvest

If people are empowered, the motive must be because they are trusted and believed to be a great resource, not because employers want to save money on supervision. If unsafe acts are emphasized, the motive must be because managers want to teach people so they don't depreciate their lives, not because they want to find someone to blame.

HOW CAN WE BENEFIT? WHAT *SHOULD* WE DO?

Tell People What Is in It for Them

Help people see safety through a wide angle lens with surround sound, not through a telescope with a narrow field of vision. When they see the benefits and believe in the virtues of conservation instead of waste, they'll do the right thing. You won't have to do *much* listing of critical behaviors and providing soon, certain, and positive reinforcement. The right attitudes, behaviors, and outcomes will be imbedded in their souls.

Abandon Efforts that Focus on Inspection of Physical Features

This is like deleting the quality control inspectors from the production line. Each assembler is a quality control inspector. Same concept here.

Appeal to Their Best Instincts

Managers tend to appeal to the most base instincts of people. They think they must give them prizes to get them to work hard or safely. That's not true. You will get out of people about what you expect of them.

About fifteen years ago, I gathered my first little league team for its first practice. I had played Division I baseball with many athletes who later played in the big leagues and was therefore confident in my game knowledge. But, I was a little intimidated by this group of ten to twelve year olds that included my eleven year old son. I felt like I was on trial in front of his friends and teammates.

I had a great idea to break the ice. I asked them to tell me something about themselves. Most of the comments were what you would expect. I live here or there or I like football. One of the boys took a risk. David Stevenson said, "I like girls." That's high risk for an eleven year old boy among peers; at least it was fifteen years ago. I took an instant liking to him. I treated him with respect. We bonded almost immediately.

Later, I found out from his parents that his experience on our team was a highlight in his life because he was constantly in trouble in school. He had even gotten into physical skirmishes with teachers. You see, David was very big for his age and didn't like to be pushed around or treated unfairly. When he was, he reacted. When he stepped out of line, all I had to do was look at him; he adjusted immediately. In all honesty, all this occurred without any real thought or intent on my part. I learned from it on reflection. I liked David. I showed him respect. I expected a lot of him. He trusted me. He knew I liked him and was interested in him. He wanted to live up to my expectations of him, so he knocked himself out. I think David was very normal.

Most people will deliver what you expect of them if they trust you and think you sincerely care about them.

Teach Them What They Need to Know

A year later we won the league championship. I don't think we had the best players because I didn't know the boys very well when the selection occurred. I did teach them what they needed to know to win though. Then I provided a lot of encouragement. There was very little criticism, and when there was, it was quickly followed by renewed concern and support. This is a formula that works with people. We need to define the tasks, making sure all the necessary skills are taught and understood. Then, we must provide enough time to practice. Finally, there must be successes and encouragement. It's like an athletic team. But you can't just put the same uniform on everyone and assume you have a team. We do that all too frequently in the work environment.

Reward Them for Doing the Right Thing

This should be in the form of pats on the back and intangible rewards. It need not and should not take the form of sizable awards. Think about when you have felt the best about recognition. Was it when you got some prize or when someone you respected acknowledged the skill and/or dedication associated with something you had accomplished?

Perform Accident Investigations the Right Way for the Right Reasons. Don't Stop Until You Have Found the Root Cause(s)

Never blame the people involved in the incident. *Always* blame the process and management. With very few exceptions (I can't think of any off-hand) you'll always be right. And even if you're not, just think how good the employees will feel about you. A few will take advantage of you. All the rest (and there will be a lot of them) will rally for your interests. They'll do much better at protecting you than you could do for yourself.

This chapter hasn't given you a recipe for success, but I hope it has given you some ideas and cause for introspection. Few one-size-fits-all answers work when dealing with people, but there are some things that almost never work and others that work well most of the time. Since humanity is such a powerful resource, you can't stop trying to get better at how to optimize it. Accident prevention is a good place to start.

9

PUT THE RIGHT PERSON IN CHARGE

"A good leader is a good follower."

Just as a space craft cannot perform beyond the designs of its engineers, an organization's safety and loss prevention results cannot rise above the quality of an organization's safety practitioners, safety advocates, and loss prevention leadership. The reasons seem quite obvious. So I am puzzled when management wants better safety performance but resists doing anything about their professional guidance on the subject. Perhaps part of the problem is that few people know what characterizes an excellent safety professional. The focus of this chapter, then, will be to paint a picture of the ideal safety professionals and where they should be positioned in an organization.

Two premises that underlie the treatment of this subject:

Premise #1 Leadership is critical to the success of any activity.

Premise #2 Safety is no different than any other activity; therefore, the safety professional must be a leader.

WHAT SHOULD THE BOSS LOOK LIKE?

Over the years, I have had the opportunity to speak to many groups of safety professionals. A frequent question is, "To whom should the safety professional report?"

I start my answer by saying that the title, status, or salary of your boss do not matter. The arguments often center around whether you can be effective if you report to the human resources professional; or if reporting to the corporate counsel focuses to much attention on regulation; or if reporting to operations creates a conflict of interest that is hard to breach; and so on. My position is that only the criteria below matter. Explanations follow.

Criteria #1 The bosses should understand what you do and be empathetic with the obstacles you face. They should have a sincere interest in loss prevention.

Criteria #2 The bosses should have stature (not status) in the company and should be among the most influential people at their level. The bosses should be willing to use their influence on your behalf or on behalf of the loss prevention function whenever and wherever needed.

Criteria #1 Explained

In 1981, my good friend Chuck Culbertson wrote a booklet, *Managing Your Safety Manager*, printed by The Risk and Insurance Management Society (RIMS). Every person who manages a safety manager should read this booklet or somehow get the information it covers.

Chuck starts with a short section on the history of safety. I never liked history in school because I had trouble remembering names and seeing its relevancy. Now, I love history and enjoy it more every day. Why? Because I realize that I can avoid a lot of mistakes by learning from other peoples' experiences without personally incurring the pain they may have suffered. The bosses should understand a little about the history of the safety movement—where it is now and how it got there. Knowing a little about the asbestos scare and scam, for example, could help avoid a similar debacle. (Chapter 13 talks more about asbestos and emotionalism in safety.)

The bosses should know when they need safety staff and how to hire competent people. That is easily said but very hard to do. Selecting and hiring safety professionals will be covered in detail later in the chapter. As a related matter, they should know what goes into the safety budget or what the safety staff needs in the way of support.

A little story will help to illustrate my point here.

When money gets tight and budgets need to be pared down, there are always some favorite targets. Travel and entertainment are seen as soft items that are not directly related to revenues and profits, so they come under early and strong attack. The first time that happened to me, my instincts served me very well. My boss called me in and told me that my department travel budget was going to be cut twenty percent. I had twenty professionals on my staff. I responded quickly with a question. I asked him whom I ought to fire and how soon.

He quickly explained that I didn't understand. We didn't have to trim staff; we only needed to trim the travel budget. I explained that the two were tied together. Our effectiveness was based on our presence at the sites. We only prepared for the visits and wrote reports in the office. The real action took place in the field, and if we couldn't go, we were of no use to the company and so should not be collecting our salaries under false pretences. He looked quizzical and changed the subject. During our next meeting, he told me that he found a way to avoid cutting our travel budget.

I found out how some time later; others in my division took hits larger than their fair share.

The leader of company loss prevention must understand the nature of the environmental safety and health person's relationship with the people with whom they interface. Particular emphasis is placed on the relationship with operations, since that is who has most of the accidents and who is, effectively, the client in most companies or organizations. The uniqueness in that relationship must be recognized and understood. I used to tell people who worked for me that if the plant managers felt they needed us, we'd always be secure in our positions. If only the CEO liked us, we could still become a luxury he couldn't afford, and during hard times we could find ourselves unemployed. It's the old, you-need-to-know-who-the-customer-is drill. Through numerous belt tightenings, I never lost a position or person.

The bosses must also understand how safety is measured, including its vagaries, as discussed several times in this book. Results are what the bosses will need to report to their leaders, so understanding their significance, strengths, and weaknesses is vital.

Finally, what about the bosses' level of commitment? Safety professionals seem to have a particular affinity for what they do. They really believe protecting people, property, and the environment is important. If their leader is less than enthusiastic in spite of what they may know, it will take the "umph" out of the whole process.

Criteria #2 Explained

Why must the boss be a leader among leaders? Because safety is harder to manage than most other functions in an organization. That's the only answer.

At a plant level, the plant manager is clearly in charge. Anyone who reports to this person, has their complete confidence, and routinely serves as their proxy could meet the qualifications for being the boss, although there is more than one company with a "world-class" safety effort that insists on the safety person reporting directly to the plant managers. That may not be practical in some industries and at very large facilities.

At group, division, and corporate levels, organization structure gets a lot more complicated. Complex or not, people tend to know who among equals (i.e., on the same corporate level) seems to be more equal than their colleagues. That is the person to whom loss prevention should report. This is like when a Supreme Court Justice (in the "community standards" opinion) said of pornography, "I can't define it but I know it when I see it." In any organization, people may not be able to explain why some associates get their way more often than others, but they know who they are. It's not always the person with the biggest office either. If you relate to that, I'm done; if you don't, I probably can't explain it so you will understand; so I'm done anyway.

Moving to the second part of this criteria, bosses must use their influence on behalf of safety. Someone can theoretically understand a subject and be powerful, but if they are not *really* committed to the mission, all that may be for naught.

Remember Dick Scheifele, the vice chairman I mentioned in Chapter 5? He was always an advocate and for a while my boss's boss. When I needed to hire a claims manager, we couldn't get the job graded high enough to attract the person we knew we needed. He took it right to the Chairman, Ralph Ward, and spent some of his influence. And we got our way. When I talk later about the importance of attracting the best people, I'll elaborate on the outcome of his efforts on my behalf and safety at Chesebrough-Ponds.

Let me expose my **executive "bullet theory"** here. It postulates that every person of any influence in an organization is apportioned a certain number of bullets at the beginning of each year. The number varies (based on rank) but not significantly. What does vary is the size of the projectile and its impact potential. Those at the bottom of the authority pyramid get 22's. Those at the top get howitzer shells; their effect can be huge, but precise aiming is not easy.

This theory came to me one day when I was in Ralph Ward's office and he received a phone call. He immediately gave me a confusing assignment. I was to use my influence to attempt to achieve an attitude adjustment on someone who was my senior in the organization. I was puzzled because he knew it would take me a lot of time to try and my success would be doubtful. It was important to him or he would not have counseled me on what needed to be done and how I might approach it. One quick phone call from him to the person would have immediately and dramatically changed their attitude.

That was just the point that took me a while to assimilate and what led to my bullet theory. If he had made the call, he would have used a howitzer to hit a target that could have been serviced by a much smaller projectile. When it hit, it would have impacted a much larger target than necessary, possibly causing harmful reverberations. At that moment, I realized that I had no ambition to be a CEO or anything like it.

Being a CEO is lonely. Everyone thinks you can do anything when reality is that you can do big things but often are powerless to deal with little things. If I saw a stain on an office carpet, I could call office services and get it cleaned up. If I was the CEO and was aware of my influence, I'd know that if I called, all the carpets in that building and in all facilities I was scheduled to visit within the next month would be cleaned, no matter what the cost. Seriously soiled carpets may even be replaced. The call could unwittingly cost the company tens of thousands of dollars.

I'll never forget when we visited the Rochester Ragu Plant when they won the Chairman's Safety Award. Those visits always included not only the award of the inscribed bowl and gifts for the employees but a plant tour. I was just behind the plant manager who was touring the Chairman and overheard Ralph inquire about the floor in the warehouse. It was beautiful and immaculately clean. He innocently asked when it had been painted. The sheepish answer was the day before. I say he innocently asked because he quickly went on when he realized that he had embarrassed the plant manager and should have known the answer.

The point here is that the leader must be willing to use some of their bullets and master timing and aim. The safety function needs to report to the person who has the right size bullets. In my example, Dick knew how important that claims manager was to saving us a lot of money, and he knew Ralph knew it too. He used one of his big bullets at the right time.

Before we move on, let's look at examples of the wrong kind of leaders.

The "Play It Safe" Manager

This type of manager will not be up to the demands placed on the leader of protecting company assets. Often, the "book" doesn't apply. Boldness is required.

The Insecure Manager

Insecure people exemplify those who Covey suggests have a Scarcity Mentality, which is the antithesis of the Abundance Mentality needed to encourage a Win/Win orientation. Safety is easy to rationalize; it is easy to bully; it is even relatively easy to forget. Only Win/Win will drive safety. The insecure manager can't think that way.

The Self-Centered Manager

If loss prevention is a people business, and everyone knows it is, a leader who is people insensitive will gravitate to positions that lead to failures in protecting people, property, and the environment.

The Egotistical Manager

They "know" everything. Nobody knows everything about safety. Like medicine, it is more art than science. Most of the formulas for success are empirical. They have been learned by repeated experiences. Egotists have trouble learning from experiences.

The Ethically Disoriented Manager

This person may work well with some people but will be like oil and water with the usually altruistic safety practitioner, as mentioned before. Since the boss almost always wins, exit the safety person. This may be one very inefficient, resource consuming, and sadistic way to deal with competency problems.

There are other fatal flaws but this should provide some ideas of what to avoid. Extricating yourself from one of these situations can be very complex and is beyond the scope of this chapter.

WHAT SHOULD THE SAFETY PROFESSIONAL BE LIKE?

The safety professional should be adequate to the task. There is no one-size-fits-all description that can be given here, but there are many guidelines that offer direction.

Bill Driscoll was the Vice President for Safety at OxyChem, one of the most enlightened companies in the world when it comes to safety. My relationship with Bill goes back to the beginnings of both of our careers in the late 1960s in New Jersey. When I went into the consulting business, Bill was my first client and one of my best, if not *the* best. Bill knew what he wanted and he was pretty sure how to get it. He also had the support of his management because he has the *essential characteristic* of an effective loss prevention leader. *He is credible.*

Years earlier, while working for Jack Snyder, Corporate Safety Director at Merck, one experience helped me understand the critical role communication skills play in establishing credibility. Jack and I were discussing the filling of a critical safety position in the organization. His bias (and that of the organization) was toward technical qualifications. I focused on credibility rather than technical know how. On reflection, the safety professionals who made the greatest contributions at Merck were not those who had technical credentials but those who could sell themselves and their ideas, i.e., those who could communicate well.

I brought this experience to my work with OxyChem, where Bill Driscoll wanted to facilitate the movement from having a good safety result to being the best they could be. At our first meeting, he suggested that he probably couldn't do that with the people they had in place. He would need to raise the bar for safety professional performance. How will a high jumper know how high to jump if there is no bar? And how can they get better if the bar isn't periodically raised?

A rigorous screening process followed. It involved resume review, group interviews, unique reference checks, and closure that brought seventeen of the best people in the safety business into the OxyChem organization.

Another critical element occurred just slightly behind hiring the best new personnel. We decided exactly what skills and abilities the optimal safety resource person or internal consultant should have. We defined the core competencies and developed a twenty-five page booklet, *Planning for Performance,* that did everything from define the competencies to provide the mechanics of how to use the development guide to facilitate developing the best safety professionals in the business. I helped define the competencies; numerous people in the organization provided input; and a human resources consultant formatted the booklet and made it user-friendly and compatible with the company performance appraisal system, although it was distinctly separate and apart from that process.

The booklet would seem like a product; it was not. The product was the *process* of generating and administrating the *apparent* product. Other staff groups in the company were so impressed with the process that they tried to mimic it in several ways. Safety was leading the way at OxyChem as it can in any organization if there is credible and competent leadership.

The competencies that OxyChem developed, broken into three categories, are listed on page 154.

This development tool is very useful and has numerous hidden benefits. It lets everyone know what a safety professional should know. By its very nature, it enhances and clarifies the vision that people have of the safety profession.

This discussion of safety professional competencies reminds me of a conversation I had with Wayne Christensen when he was the executive director of ASSE in the mid 1970s. I advocated that the Society do something like this definition of the profession. Wayne pointed out that few members would measure up very well. I said that was exactly the point; here was an opportunity to change that. He used the light bulb analogy suggesting that if management saw how much light a 200 watt bulb could provide and how nice it would be and realized that they had a 15 watt bulb, the first thing they may do is go out and change the bulb. I said, "So what." I hoped that raising expectations could raise performance and the perceived value of safety professionals.

OxyChem's Competencies

I. Interaction

1. Communications - Oral
2. Communications - Verbal
3. Employee Development - Resource Planning and Needs
4. Employee Development - Training and Measurement
5. Influence and Persuasion
6. Leadership
7. Meeting Skills

II. Administration

8. Computer Skills
9. Economics of Safety
10. Insurance
11. Organization and Record Keeping
12. Planning and Objective Setting
13. Problem Identification and Solving
14. Public Affairs
15. Safety Procedures
16. Security Management
17. Time Management

III. Technical

18. Regulatory Expertise
19. Process Equipment
20. Ergonomics
21. Industrial Hygiene
22. Occupational Health
23. Preventive Maintenance
24. Process Safety Management
25. Property Protection Systems
26. Emergency Response

With a field as broad as loss prevention, the practitioner's greatest asset may be knowing what they don't know and knowing where to find help when needed. My greatest success was in finding and motivating outstanding people who made me look good. It wasn't always easy, but it was always worth the effort. Support people with specific expertise are essential resources for the safety professional. The ideal safety professional should have the skills to coordinate and consult as needed and then be able to integrate the support.

FINDING HELP

Probably the most obvious discipline in which safety professionals would be unlikely to have the necessary in-depth expertise is medicine. I realized that early on and used my network of friends to find a consulting board certified occupational physician, Dr. Howard "Buzz" Sawyer. About the third year I was using him, Ralph cut the $50,000 I had wanted for the consultant. I appealed to Bob Bennett, who was then my division head, and he suggested I appeal to Ralph. He knew that we had a physician in the R & D area whom I could access. He was a pediatrician who presided over the cosmetics testing we did. I explained to Ralph the importance of certification in occupational medicine and to emphasize, asked him if his son had a brain tumor if he would like the surgery done by a pediatrician or a neurosurgeon. The $50,000 was restored.

For senior executives, at least one incident clearly justified that and all other expenditures for consulting occupational physicians. We had a plant that produced bandages. We treated the gauze with slightly unpleasant but not exotic chemicals. One of the employees became ill with renal failure, which caused permanent, irreversible damage to his kidneys. He would probably be permanently disabled and perhaps even unable to return to work.

He made a workers' compensation claim that would have reached multiple six figures. There was talk of other lawsuits for negligence by us and the suppliers of the chemicals. I doubted any relationship between our plant exposures and his health problems and authorized Buzz to spend whatever time was necessary to research the matter.

His investigation was fruitful and saved us a lot of money. Our employee had experienced heart arrhythmias and was taking medication for it. He subsequently also had an infection and was given an antibiotic for that. Both drugs could, in time lower white blood cell count. Together the effect was devastating. His resistance was reduced and then the kidney problem ensued. Chemical exposure was clearly not his problem. When Buzz *explained* what he had found to the cardiologist and the internist, they encouraged our employee to drop his claims against us, which he did.

As our occupational medicine needs expanded, it became apparent that we needed to hire our own occupational physician. I first looked at the doctors we had servicing our plants. Dr. Bob Caven was doing an exceptional job in our Bangor plant, in spite of very few resources, little support and with no reinforcements in sight. I hired him and one episode made me certain I had the right man.

Early in our efforts to abate losses due to soft tissue injuries, I discovered a chiropractor who could be an effective part of our team. I sensed that he could also help our plant doctors. I arranged a seminar for our doctors, and my new consulting chiropractor was to be one of the key speakers. Several of the M.D.s were reluctant to attend a seminar with a chiropractor speaking. I added a prominent physiatrist, an M.D. trained to deal with musculoskeltal afflictions and rehabilitation issues.

Bob Caven and some of the other M.D.s still balked. I gave them a choice—they could come and participate or abandon their relationship with us. They all came but didn't look happy when they arrived. The chiropractor was ignored...until *after* his talk. It was quickly apparent that he knew more about certain salient aspects of our employee's afflictions than anyone else in the room. The M.D.'s argued over who would sit next to him at lunch. Bob came over to me and apologized for his initial reaction to my idea. I knew he was my man because he was flexible and open-minded enough to meet the challenges we were to face. One of our greatest was in the area of soft tissue injuries. He did an exceptional job.

We got victimized by soft tissue injuries in the early 1980s. Before everyone else got on the bandwagon, I couldn't find anyone, including M.D.s, who seemed to have answers.

One evening I was visiting a family from my church. I didn't know them well before this visit. I was there following a weekend playing basketball that had left my body reminding me that I was nearing forty and had abused my body too often in the past. One hip was so sore that I couldn't lift my leg when sitting without help from my upper extremities. Sally, the lady of the house, noticed my discomfort and mentioned that her husband was a chiropractor and could "fix" my problem. I'd had the pain many times before and was well aware of its solution/resolution. It would get better each day and in two to four weeks would leave me fairly functional.

I wasn't exactly sure what a chiropractor was but the name elicited feelings of apprehension. I suggested that my visit was not designed to harvest free professional services. Sally persisted, and, as if on cue, Pat, her husband, entered the room. She quickly told him of my problem and he suggested I lay down on the floor. What do I do now? I had not pre-thought this; it happened too fast. My mind was racing—how could I avoid further damage to my already impaired ability to move? Then a powerful thought overcame me. How could this be made any worse as long as he didn't have a knife? I was trapped and reluctantly submitted.

Pat asked me a little bit about the history of my infirmity. He went right to the spot and pressed it. Tears came to my eyes as he didn't even have to ask if he had struck the source of my discomfort. He told me that he was going to press very hard and that the initial pain would exceed anything that I had experienced up to that time and then, it would gradually abate. Sure enough, he hurt me. But remarkably, the pain then dissipated. I was then able to get up, walk, and sit with no pain or discomfort. I sought an explanation. I got a short course in anatomy and learned what got me to where I was, exactly what he did, why it worked, and what the future held for me and my hip. It all made sense.

I explained our soft tissue injury problems at Chesebrough-Ponds to my new-found chiropractor friend and deliverer from pain, Dr. Pat Venditti. He wanted to visit our factories and see how our employees were getting hurt. He felt he cured people and they just kept coming back. He wanted to see what caused the injuries. I told Ralph about the potential I thought Pat had to help us. Ralph's biases led him to suggest that we probably shouldn't be dealing with a chiropractor. Was I glad Dick was there. Not many people would have done what he did then. He asked if Ralph had ever been treated by a chiropractor. Ralph said no and Dick said, "Then what do you know about them?" Through his leadership and courage, we were able to tap a great asset. Pat became a critical part of solving our complex problem.

We opted for logic instead of emotion. See Chapter 13 for a further discussion of that subject.

When untoward events do occur, after-the-fact loss prevention (i.e., damage control) becomes necessary and really amounts to claims handling. It is generally done poorly. In the claims handling function, you can often save more money quicker than in almost any other area of business or loss prevention. This fact is a well-kept secret in industry and only a few companies, like Marriott and a handful of self-insureds, know it and act on it. I learned the secret from my friend Chuck Culbertson and then researched and documented what he had told me. Chesebrough-Ponds had inexperienced clerks authorizing settlements for over $50,000 on our behalf that, with mark-ups, could cost us up to six figures. The most senior officers in our company didn't have that kind of authority. Needless to say, we found our service providers (third party administrators) were making a lot of mistakes on our behalf.

When I asked Ralph for authorization to hire a claims manager, I was armed to the hilt with justification. His only response was, "Why hasn't anyone ever asked me for this before?" Again, he was way ahead of us. I refused to answer the question and he didn't press me. He probably knew exactly why not and even who was responsible.

We hired George Nelson, and on some days he saved his annual compensation in a few hours with a settlement that was a fraction of the reserves for a case. I used to send copies of those to my new boss, Ed Hughes, with a little note explaining just what had happened and referencing it to what we paid George. George became one of the most respected and appreciated people in our company. I learned a lot about claims handling but mostly to reverence people who had a lot of experience and savvy in that area. You see, no schools give degrees in claims handling.

After an injury and after the first efforts to mitigate it don't quite work, guess what comes next? You guessed it—litigation.

If they have a lawyer, you need one. As with a company's CEO, if a lawyer is good, you can't pay him too much. If he is bad, anything you pay is too much. Unfortunately, when I looked at how Chesebrough was litigating cases in Maine, I concluded that we almost always lost, so I could have sent my teenage daughters and they could have done as well. Come to think of it, at least one of them could have done better.

I tried to find the person who handled claims best in Maine and was directed to the Fred S. James office where I met Joe McCarthy. I gave Joe the ultimate compliment when I came to know how he served his clients and particularly how he litigated claims in a very difficult atmosphere. I said, "You don't behave or look like an insurance broker." He tried to return the favor by suggesting, "You don't look or behave like a safety engineer." Actually, I probably don't but I like most things about safety people and so have never been ashamed at being identified with them.

I asked Joe who was the best compensation attorney in Maine. He referred me to a person that I would be unlikely to uncover using normal procedures. John Lambert was a young attorney working for a small law firm. John didn't litigate many cases but the comp savvy people in Maine knew he was the best. According to Joe's description, John wore wash and wear shirts that looked like he had slept in them, wore rubber boots for even longer than needed in Portland, Maine, and always looked like he needed his hair cut or combed or both. The rest of his physical description was likewise disparate with his capabilities.

I couldn't wait to meet John. I was in his office that very afternoon. Joe's description was perfect. I asked John just one question: "How do you prepare for and litigate a workers' compensation case?" John started by indicating that his litigation started before the case by getting to know the client and their workplace. He then went through an impressive description of preparation, strategy development, and tenacious prosecution of his case. I had on good authority that our current firm employed by our third party administrator often grabbed the file on the way to court and read it while waiting to be called.

I immediately hired John. I didn't have the authority but had already become a firm believer in what I told everyone who worked for me: "It is easier to apologize than to get permission." In this case, it turned out that I wasn't quite right. I'll tell you more about that in Chapter 13 when discussing being guided by logic instead of emotion. But, I will tell you the end of the story now. John did win a lot of cases for us, got to be well known, and went out on his own and did very well. Later, he told me that he saw that interview that I had with him as being a turning point in his career. One of our executives once asked me if I could explain John's success. He wanted a short response so I gave him one: "He hasn't been around long enough to know that he is supposed to lose."

Looking down the list of competencies, you will rapidly recognize that these four areas (medicine, chiropractic care, claims handling, and litigation) are not the only fields for which a safety professional may need to employ outside resources. They need the knowledge to get the best in *any* area needed and the courage and conviction to sell the decision makers. The right boss will help them do it.

SUMMARY

In conclusion, companies should realize that where their interests are, there also will their money be. If you see a well-supported loss prevention function, the company cares about protecting and conserving their human and physical resources; if you don't see the support, they don't *really* care about it. Visit an insurance company or broker some day. Remember Tobias's *Invisible Bankers*? You'll find the investment people in the corner offices furnished with antique reproductions. The producers (sales, marketing, and account management) will have large offices with wood desks. The underwriters will have offices, but not quite as nice. The safety engineers will have metal desks and be in cubicles. . . if they are lucky. The claims handlers will be sharing phones and have piles of folders on their desk and all over the floor. Priorities are obvious. There is nothing wrong with this pecking order, given their mission ("invisible bankers"). What is wrong is when the client doesn't understand the priorities and acts accordingly. Service users (the company) need to put the right person in charge of safety and support them so their priorities take precedence over those of the insurance industry.

10

USE A YARDSTICK EVERYONE CAN READ

"Don't measure my corn by your bushel."

Measurements of safety performance are bogus. At least how they're usually done. Yet measurement is important. If a company can't measure it, we can't manage it. I'll begin by relating a few personal stories meant to give perspective to measuring. Then, let's discuss how to do it correctly in safety/loss prevention.

A couple of years ago I moved into my sixth decade and found, as many people in that fix do, my clothes were shrinking faster than ever before. Nordic Track seemed to meet my needs. I bought one and started using it. I liked a lot of things about it, especially that it didn't hurt. (Walking, running, and cycling all left my skeletal system somewhat the worse for the benefit my cardiovascular system was enjoying.) I loved it and felt better than ever. And I lost almost twenty pounds in less than a year. (Don't believe the projections in the ads.) My model has a computer that tells you everything from your heart rate to the number of calories you are burning. It also measures distance in miles or kilometers. Every day, I recorded my progress.

One day, the batteries in the computer went dead. I couldn't drive myself to continue, for I wouldn't know how many calories I had burned and therefore couldn't record personal progress. I began reasoning; I reminded myself that I could estimate my progress since I could still tell time and knew within a few calories how much I burned per minute and per hour. I told myself that what was important was exercising, not recordkeeping. Eventually, I got mad at myself for being so ridiculous, even calling myself names. Still, I couldn't force my body to continue. The next day, with the new batteries in, I was off again.

*By pushing myself and setting little goals, I would do more each week. Without the goals and the **measuremen**t, I don't think I could have accomplished what I did.*

Then something interesting happened. I graduated. After about fifteen months, the computer didn't matter to me any more. Just knowing that I was doing the right thing and feeling better when I was doing it was enough to motivate me.

I don't think I'm much different than most other people. Sure, measurement is more important to some people than others, probably because some people enjoy the process itself or appreciate the extended benefits of exercise or other self improvement efforts. But we are all remarkably alike. We need safety measurement systems that tell us what we have done, and more importantly, motivate us. Hopefully, the mountain of extended benefits of safety will eventually overwhelm reliance on measurement as the motivation.

Sometime ago, I coached a boys' church basketball team. We were playing a championship game at the opposition's court. They arranged for scorekeepers. They were favored and had more and better players. They had beaten us during the regular season. They kept score both in a scorebook and on a big blackboard. We were behind for most of the game and then something happened. My high school age young men forgot that they were overmatched. We started to play better than I ever thought we could. The other team got scared. I kept fueling the fire. Occasionally, I'd look at the blackboard and note that the score didn't seem to be keeping pace with the dominance of our play.

Once, the other team and home grown officials raised some questions that reflected their partisanship. We checked the scorebook and found it inconsistent with the chalkboard score, and we lost a couple of points. Near the end, I was not only riveted on the game but on the clock and the chalkboard. At the end of the game, we made one final thrust and miraculously won by a point. We had a great celebration. Our opponents started checking the official scorebook. Guess what? The chalkboard was wrong. We had actually lost by one point. Interestingly, my team was hardly affected. They felt like they had done what they had to and, in their eyes, had won no matter what the scorebook said. And equally interesting, the other team didn't celebrate. They felt and still looked like losers.

We headed home "triumphant." But that wasn't the end of the story. The dad of one of our boys had videotaped the entire game and after each score showed the chalkboard. His son, our captain, watched the whole game when he got home. He called me and reported that we had actually won by ten points, not one. I was irate. My boys, who had played so hard, had been cheated either by intention or incompetence. I suggested we get redress. They wisely suggested that it didn't matter. We knew who won and what we had accomplished; what else mattered? If they knew and felt badly, they would have to live with that. If they thought they had won, so what?

So, scorekeeping is important, and it should be done fairly and accurately. But, it isn't the most important thing.

Let's take a look at what we want in a safety measurement system and then see how the current systems stack up against what we need.

What We Want a Measurement System to Be

It should be:

- Timely
- Accurate
- Intelligible
- Fair
- Consistent & Reliable
- Edifying
- Verifiable
- Personal
- Predictive

This illustrates my want list and each element is explained below. Next the systems we use are compared to what we want.

Finally, a better approach is suggested—something that represents a yardstick that we can read...and understand.

Timeliness

I don't want to find out I lost the game after I'm at home taking my shower. I want to be able to react to the score. I need real time information so I can learn from mistakes and make adjustments during the game.

Accuracy

If the score is going to be inaccurate, I'd rather not know it at all. That not only doesn't inform me but could send me in exactly the wrong direction. If I think I'm losing when I'm really winning, I could expend energy that could and should be applied at another time. If I'm making progress and mistakenly think I'm losing ground, I could abandon an effective strategy.

Intelligibility

Any system I can't understand is useless to me. If it is in a foreign language that I don't speak, I'll either ignore it or misinterpret it. Have you ever started playing a new game and been unaware of certain rules, later finding out that your approach was just the opposite of what was needed to win? If you're playing a game, it can be amusing or even funny. If someone's well-being, perhaps yours, is at stake, the humor may be hard to appreciate.

Fairness

In my basketball story, we felt okay because not that much was at stake, and we were satisfied with what we had accomplished. Our play exceeded our expectations. But, suppose the opposition had suggested that we go back each week and play with the same officials and scorekeepers. I don't think we would have been quite so forgiving or anxious to participate, knowing from the outset about the bias/incompetence of officials and scorekeepers. If measurement is unfair or even perceived as unfair, we become disenchanted very quickly.

Consistency and Reliability

If we can't replicate results, whether in a scientific experiment or in records associated with particular performance, conclusions are very confusing and a big turn off. Have you ever watched an event with subjective scoring, like ice skating or gymnastics, and felt that the judges were capricious and inconsistent? Suppose that was you being judged and it kept happening. That kills incentive, doesn't it?

The quickest way to turn someone off is to provide different responses to the same input. Under a worst case scenario, you will drive someone crazy. I vaguely remember some discussions from my basic course in psychology about tests done on animals when erratic reinforcement is provided. Pavlov's dog salivation experiments on conditioning are a good example. The dogs quickly disorient. When I pitched, I went nuts when plate umpires kept changing their strike zone. I didn't mind if their strike zone was different from other umpires, as long as they were consistent and I knew what to expect from them. Anger is one response and withdrawal another. I think that many managers and workers have at times gotten angry and many have withdrawn from the safety measurement systems to which they have been subjected.

Edifiable

If the goal is to get a better number (isn't it always?) you must learn from the score and make changes to improve your result. In those subjective scoring events mentioned before, the judge should be able to tell the performer exactly why they were marked down and what they needed to do in order to get full credit or score. Ideally, someone should be able to show the competitor or model for them what is required for a perfect score. The performer can then learn what they are doing wrong and how they can correct themselves.

Verifiability

This is quality control. If scoring is important to me, I'd really like to have someone monitoring the scorekeeper to make sure scorekeepers are applying the agreed upon rules. If possible and practical, I'd like to be involved in that process myself or at least be satisfied that it was above reproach. If we did go back to play that basketball game again, I would have wanted a clearly impartial person watching the scorekeeping. I'm sure I would have spent more time observing the chalkboard. I even think I would have had someone on my side keeping a scorebook.

Personalization

If possible and practical, I want the scorekeeping system to consider me and my circumstances. On my Nordic Track computer, I can put in my age, weight, sex, and the resistance I set on the skis. My wife uses it too. A few times I inadvertently began exercising using her settings, which are lower for weight and resistance. As soon as I realized why my readings were so low on calories burned—why I wasn't getting full credit?—I quickly changed the setting. I wanted full credit for what I was doing. I didn't want to be measured on the basis of *her* setting. I think most people react the same way.

Predictability

In a best case scenario, a measurement system will predict what we can expect to happen in the future. This is the whole aim of empirically derived mathematical equations. They define measurements of nature and then allow use to predict nature in the future. Most valid safety measurement systems offer clues on trends. But wouldn't it be even better if the measurement could predict performance more precisely, and three, six, nine months, or even a year in the future? Adjustments could be made to improve performance. Or at least we could prepare to budget and respond to what we know will happen to us.

CURRENT SYSTEMS IN COMMON USE

Three major systems are commonly used to measure loss prevention performance. Several other systems are emerging, but in most cases have seen very little practical application. Since measurement is a central issue and benefit of VIRO, several observations on these varied approaches should be noted. The table on page 173 rates each approach according to the desired criteria, from excellent (4) to useless (0).

THREE POPULAR SYSTEMS

I. Incidence Rates

Incidence rates are now usually computed using the OSHA system. The chemical industry was about the last major holdout and now they, too, have mostly converted to the OSHA system. They threatened to keep "two sets of books," but most of them soon relented, so few still use the old ANSI system. The basis for both OSHA and ANSI are recordable and lost time injuries. The rules for counting under OSHA are extensive but not particularly onerous. Over the last decade, consistency has improved considerably, particularly within any given organization. Consistency suffers when comparisons are made across industries and even within a company with significant demographic differences.

II. Workers' Compensation Costs

The number of systems for measuring workers' compensation cost experience boggles the mind. One almost universal measure is the experience modification factor that seeks to place different population groups on an even playing field by categorizing them by exposures/industries. A one (1.0) indicates an average performer within a classification. A two (2.0) means that the company/organization/group will pay twice the average and a one-half (0.5) means that costs for workers' compensation will be half the average for the industry or exposure group being evaluated. There are enough potential fog factors, such as state funds and self-insured programs of various configurations, to defy explanation in less than a complete text.

One virtue of this measure/system is that it quantifies in units that everyone is familiar with: dollars. The absolute measure can and perhaps should be the amount for which the company writes out checks during any calendar year. The amount is based on experience that goes back four years and covers a three year period. It is affected by overhead costs and administrative costs. The propensity for humans to add complexity to everything they encounter is exemplified by the workers' compensation systems.

One wise friend deals in buying and selling companies. He said that when evaluating a company, most people place a lot of credence on their balance sheet, sales, and profits. He said that he was only interested in looking at one thing: their checkbook. He maintained that all the other business success measures, including tax returns, could be fudged to confuse. There is not much you can do to your check book. You have to show your deposits and disbursements. So, in spite of all the nuances endemic to the workers' compensations systems, one can tell a lot from the checks written each year. That *is* the cost.

III. Audit Programs

Audit programs take a wide variety of forms. Some focus on compliance with regulations. Others look at the systems used to predict the desired results. Some attempt to do both. A few of the more enlightened approaches include behavioral aspects too. Most do not. One of the systems that enjoys broad based use is the so-called "Five Star System" marketed by The International Loss Control Institute (ILCI). The management consulting firm Arthur D. Little also has an environmental safety and health audit that they market. There are probably others that are not included here. Most companies have audits that are developed internally. They range from little better than the old safety inspections to very sophisticated and effective efforts. Even the

EVALUATING SAFETY MEASUREMENT SYSTEMS

Excellent = 4 Good = 3 Fair = 2 Poor = 1 Useless = 0

What We Want In A Measurement System	Injury Incidence Rates	Workers' Compensation Costs	Audits and Evaluations	Behavior Based Systems	Value Inspired Resource Optimization
Timely	1	1	2	2	4
Accurate	1	1	2	2	3
Intelligible	1	0	2	1	4
Fair	1	1	2	1	3
Consistent & Reliable	2	2	2	1	3
Edifying	1	0	2	2	4
Verifyable	2	3	2	1	4
Personal	0	2	2	2	4
Predictive	2	1	1	1	4
Total Score	11	11	17	13	32

best still smack of an "inspection" mentality and only pay token attention to the *real* causes of mishaps.

OTHER SYSTEMS

Behavior-Based System Measurement. The *behavior-based* system seeks to define critical behaviors and then observe and record them. That process yields indicators of future performance and focuses on what this approach sees as the root causes of bad outcomes. It provides the mechanism to reinforce desired behaviors and extinguish behaviors that predispose and predict accidents. The technique involves job observation to count and record behaviors. This approach is only used by a fraction of organizations, but those who use it become very devoted. It has a history of more success in encouraging better performance than the older, more traditional measurement techniques.

This text won't discuss in detail any other measurement systems because they are so obscure that only the intelligentsia of the safety profession are likely to know of them. They have very little, if any practical application, but they will be mentioned briefly as described in the text *The Measurement of Safety Performance* by William E. Terrants. Bill is an old friend who invited me to run for an office that led directly to my presidency of the ASSE. I'm very fond of him and this book was his labor of love, so I hate to criticize it. I know and respect several of the contributors. My hang-up with it is that it is difficult to understand and is hard reading for anyone outside our field. It adds to our body of knowledge (that is reason enough for its creation) but violates the basic tenet of this value, understandability. Also, it doesn't mention workers' compensation costs and therefore fails to deal with the only thing that *really* matters to for-profit organizations. . .dollars.

Measurement Based on a Structural Concept of Accident Causation

This *theory* ties statistics to management decisions and suggests four levels of causal relationships. Level one consists of accidents that are symptomatic of failures in the man-technological system that comprise the second level—the operating system. The people/materials failures reflect management inadequacies, which make up the third level. The fourth level causing failure is management's inadequate knowledge and inappropriate *values*, which lead them to misplace priorities and make poor decisions.

This theory could be seen as a first cousin to VIRO. It has not been widely discussed and even less often practiced in any form. There are offshoots of the structural concept of accident causation, but I have not seen any of them reduced to a practical or functional process. Perhaps the reasons are that: 1) its philosophical basis has not been well developed, and 2) there is no implementation strategy and tools, such as the maturity grid suggested for VIRO in Chapter 18.

Measurement of Safety Engineering Performance

This approach suggests that measuring the quality of engineering as it relates to hazard or exposure reduction will indicate performance. John V. Grimaldi, who describes this approach in Terrants' book, suggests that such an approach is more likely to affect severity than frequency. Back in November 1976, I published an article in *Professional Safety* on an Engineering Project Planner. It was a tool to engineer out hazards during the construction process. Hundreds of people have asked for complete copies of the planner, but I have only seen spotty application, and then I'm not sure it has contributed much to accident prevention. Measuring safety by measuring engineering acuity relies on the premise that accidents result from unsafe conditions and equipment. In spite of (or perhaps because of) my work on safety *engineering*, I reject that premise in favor of the belief that people cause accidents.

Measuring Safety Attitudes

Measuring safety attitudes could be viewed as a link between VIRO and behavior-based safety. Unfortunately, it has never caught on. The descriptions of the subject in works such as the chapter by Harold M. Schroder in Tarrants' book are mostly philosophical and don't delve into practical applications. The relationship between behaviors and values is not developed.

WHAT'S WRONG WITH WHAT WE HAVE?

My criticisms are many but can be reduced to three main issues:

1. Most currently used measurement systems are reactive and deal with symptoms instead of the root causes. A few approaches (e.g., Behavior-Based System) appear to have potential but that potential has largely gone unrealized, probably because they have not touched the heartstrings of management. VIRO will resonate with management because it incorporates their values and those of the workers. And, *values are the true root causes*.

2. These systems are largely unrelated to the measurement systems with which management is familiar. They are outside the comfort zone of most of the people who are forced to use them. So, they only react to them when they must. VIRO overcomes this obstacle since safety is measured by the company's values, values that underlie all the other facets of company operations.

3. Current measurement systems are not *really* understood by those who tend to get hurt and matter most, the workers. Since the workers don't understand any of the current measurements, they are not really motivated by them. Sure, they *think* they like the gifts if they work so many hours without an injury. But they are more often victims of the measurement system than beneficiaries. With VIRO, measurement is easy for the workers to interpret. They need only look at their company's value maturity, or how closely the company's behavior corresponds to their values.

WHY CAN'T WE CHANGE?

The following are some reasons safety measurement has resisted improvement.

> ### Reasons Safety Measurement Has Resisted Improvement
>
> - Change is hard.
> - Corporations (and individuals) lack courage.
> - Corporations (and individuals) lack vision.
> - Corporate politics confound efforts to change.
> - We're satisfied.

These are my perceptions. I don't hold them as inviolate. Meditate on them, even come up with your own. This is not just an exercise. It is important. If companies can't figure out why they haven't changed in the past and how to overcome those obstacles, they're condemned to eternal mediocrity by measurement dysfunction.

Change Is Hard

Once we wed to a process, it is very hard to change, unless we're traumatized or care passionately. Only a few companies or persons are passionate about safety, and even those have had little reason to correlate the few bad things that have happened to them with the way they keep records. Measuring devices are simply too obscure to make the need for change obvious.

Corporations (and Individuals) Lack Courage

Rejecting traditional measurement in favor of something unproved or generally unused takes an act of corporate boldness, if not courage. Bold and courageous people or companies are in short supply...particularly when it comes to safety.

Corporations (and Individuals) Lack Vision

Human history is not filled with stories of visionary organizations. The ones that come quickly to mind, like IBM, Standard Oil, Apple, Microsoft, Marriott, Weyerhaeuser, Ford, MADD, ACLU, McDonald's, and the LDS Church, have enjoyed growth and prosperity. For some of them, only until they lost their vision. You can decide which ones. Changing the records on which you rely will take vision. This book's job is to try to stimulate the vision.

Corporate Politics Confound Efforts to Change

Issues that don't clearly fit into one of the pieces of "Humpty Dumpty" (Remember *Reengineering the Corporation?*) create turf battles. It's easy to visualize the public affairs and human resource people's response when a safety person suggests using VIRO to enrich almost every aspect of corporate performance. You can almost always see the wheels turning and the obstacles being mentally manufactured. It will be the old "if it wasn't invented here, it can't be any good" scenario.

We're Satisfied

So was IBM. Look what happened to them. By contrast, what happened when the first McDonald's opened in Russia? What happened when Russians started to watch CNN? Their dissatisfaction drove them. No one has seen the benefits of a *truly* enlightened approach to loss prevention and keeping score of their progress. When it happens, loss prevention could well become the latest corporate fad.

HOW DO WE GET GOING?

Safety managers need to catch a vision of what a values-driven approach to protecting and optimizing resources will do for them. They need to really want to find the root causes of our dilemmas. Look at the Ruby Ridge affair and the Branch Davidian fiasco at Waco. The Congress is doing expensive investigations of the performance of government agencies. I am no fan of big government. But for crying out loud—government had very little to do with what happened there or for that matter in Oklahoma

City or the World Trade Center. Those losses were due to people with very twisted values who did very bad things. Sure, some of those losses could have been slightly mitigated by enhanced reaction. But that's a long way from the root cause.

That's not the point, though. The root causes have been left to victimize us again and focus attention on the lesser characters in the affair. Could anyone change the values of the people who were the perpetrators? Probably not. Could anyone change the cultural atmosphere within which they operated to succeed at their dastardly deeds? Of course they can. Will it be hard to do? Sure it will. Can people afford not to try? Of course they can't. The only questions still needing answers are when and how. The answer to the first question is obvious: As soon as possible. And they need to get working on the second one immediately.

They need to find someone in authority who is excited about the prospects of driving their organization on the basis of values; then they need to find facilitators inside and outside their organizations; finally, they need to resolve, not to flinch. Everyone needs to get flat out excited about each step in the process. Chapter 19 will talk a little bit about someone who leads a major corporation that appears to be onto this.

ADOPTING THE YARDSTICK EVERYONE CAN READ

This section must start with a dedication to the VIRO concept and blend it with other measurement systems that are being used. The goal should be for VIRO to become the flagship of the measurement techniques and one that will help simplify the others. At the same time, you should resolutely move toward measuring safety performance on the basis of organizational values, culture, attitudes, and behaviors rather than end results like incident rates and workers' compensation costs. Clearly focus on *process* rather than results, on indicators and predictors rather than outcomes. Our measurement system will confirm our claimed priority of process focus instead of refuting it—the case with current systems.

The yardstick of choice will be built around measuring organization values that correlate with accident-free performance, as demonstrated by enhanced workers' compensation measurement systems and accident rates. Comp costs and accident rates will be tied to VIRO with a highly developed observation of the behavior process. An audit program will insure that all is working well and place the emphasis where it belongs, on values, culture, and attitudes.

In closing this chapter, refer again to the rating of different measurement systems against my want list of nine criteria (see page 173). At this point, go ahead and rate your company's measurement system and see how it compares with this process. The ultimate yardstick will become much more tangible when the maturity grid system for assessing organization values is explained in Chapter 18.

11

SELL BENEFITS ... AND THERE ARE MANY

"Benefits, like flowers, please most when they are fresh."

. . . .

"Write injuries in dust, benefits in marble."

It was stated earlier that saving money, or more precisely, protecting what you have earned, should not drive loss prevention. Nevertheless, the primary purpose of any business is making money. Certainly, profit protection can and should be used to sell safety, but only in addition to the right reasons (see Chapter 5) and bolstered by the mountain of bonus benefits. This chapter explains the profit and extra benefits of safety that make it easy to "sell" and then offers success stories.

PROFIT PROTECTION

What would you say if you were told that loss prevention efforts represent the best way for a significant number of companies to increase their profits ten percent or more? Some of you would probably agree, but would also be hard-pressed to demonstrate it in a way that a chief financial officer would understand, believe, and want to go out and tell others. Other readers are probably thinking "no way." However, the savings potential and a lot more *can* be demonstrated.

This demonstration is based on the most credible aspects of the Risk and Insurance Management Society (RIMS) publication, *Cost•of•Risk Survey*. There is more than a little irony in the fact that RIMS's own survey is used to suggest not only that their name is a misnomer but that they are on the wrong track and, worse yet, have derailed most of corporate America's thinking. This is a corollary to the main point that will be demonstrated.

The *Cost•of•Risk Survey* is a published survey of more than four hundred companies in twenty-four industry groups averaging about ten thousand employees. It provides a treasury of facts concerning insurance and loss costs for a representative sample of business in our country. To lay the groundwork for this proof, a few of the publication's most defensible and easy to understand facts will be used.

The cost of risk is defined as the sum of :

- Net insurance premiums
- Non-reimbursed losses (self-insured, self retained)
- Risk control and loss prevention expenses
- Administrative costs

Over the years, cost of risk has consistently averaged very close to 0.50 percent of gross revenues for the companies surveyed. It has ranged from a low of 0.07 percent for the insurance industry to 2.02 percent for the "machinery" industry. This illustration will run through a hypothetical situation, and, in doing so, will use conservative numbers in all cases so the point cannot be discredited.

This example uses a $1 billion company that has an industry average cost of risk of 0.50 percent, or $5 million. The average U.S. company makes between three and four percent net profits on sales. The upper end is used so the example will produce a more modest result. Hence, assume a profit of $40 million that can be used to pay stockholder dividends, buy back stock, or finance any other purpose the management desires.

The primary assumption in this model/example is that the variance between the best and worst performers will range from one half of the average to double the average cost of risk. That needs to be defended up front because it is central to the example and the conclusions that will be reached. If you research that assumption, you will certainly find it to be *very* conservative. I have consulted for companies whose cost of risk was several times the national average for their class of industry. Placing the worst at fifty percent above average is very conservative. If "world-class" companies in safety have incidence rates of one-fifth to one-tenth the national average (and incidence rates should correlate at least roughly with the cost of risk), the one half bottom end is also very conservative. What all this means is that the upper end of the cost of risk for the sample company would be $7.5 million and the bottom—realistically achievable—would be $2.5 million.

The Example

Minimum	Average	Maximum
0.25% ($2.5M)	0.50% ($5M)	0.75%($7.5M)

This produces a $5 million best-to-worst spread ($7.5M-$2.5M=$5M) as demonstrated above. If we assume a forty percent effective tax rate, which also is very conservative, the difference between the best and worst company's profits, based on the loss prevention efficiency/results, is $3 million. That's 7.5 percent of our company's $40 million in profits. *Using the most conservative numbers, the average company profits are impacted between five and ten percent, based on whether they are incompetent or excellent at safety.* This does not take into consideration any side, intangible, or indirect benefits. Only dollars are considered; i.e., stockholder dividends, real profits or earned money that can be retained or lost through careless management.

Every member of management in every organization should be aware of facts like these. But, that $3 million of extra profits for our typical company is just the tip of the economic iceberg when safety is done well.

Recall my holiday "vacation" in my suit with Ralph Ward at Chesebrough-Ponds. He and I philosophized on this subject a little bit. We discussed one of the options to reducing accidents/costs: you just make more money. Since we had made a lot of acquisitions, we could just buy another company that made enough money to cover our excess losses. A good rule of thumb for buying companies is that they cost about what their annual revenues are. That could be easily called into question because assets, return on investment, profitability, growth potential, value of trademarks, and many other factors go into determining the cost/value of an enterprise. Probably the most important factor, though, is an enterprise's ability to generate profits.

To simplify our discussion, Ralph and I considered a typical company making about four percent profits after tax. At the forty percent effective tax rate, that translates to total profits of 6.67 percent. For every $100 million in sales, you can cover $6.67 million of excess losses. With the excess costs for workers' compensation that we were seeing at Chesebrough of between $10 and $15 million, we needed to buy a $200 million company with an average profit picture that would probably cost us about $200 million to buy. Choosing the alternative—spend $3 million to stop the hemorrhaging—didn't require a whole lot of careful consideration. If you buy a company to offset losses and fork out the $200 million, you have still not increased overall profits and have degraded your return on investment and stockholder equity, both important measures of company worth and vitality. If we look back at the $1 billion company, they need to buy a roughly $75 million company to cover their $5 million in unnecessary losses. Of course another option is just to grow the business $75 million. Is that a more pleasing prospect? (This now starts to get like a tactic that safety people have used frequently to justify what we do by showing how the cost of a product or number of sales must be raised to cover losses. A global corporate analysis is more compelling, more like looking at the world through a wide angle lens instead of a narrow width field telescope.)

The alternative of loss prevention is very persuasive. I borrowed my slogan on my stationary, "Can you afford not to do it?," from my good friend Chuck Culbertson, who worked for Bill Marriott before Chuck's premature death from cancer. It is always a good question to ask yourself. In fact, I used it effectively even in marriage counseling I did for my church service. People can always see the benefits of jettisoning a partner. They rarely see *clearly all* the disadvantages of divorce. When they carefully consider the alternative, they often see their present situation in a new light.

The point could be raised that this discussion has considered extremes and most companies are somewhere in the middle. That may be true, but all companies certainly have the potential to be at one of the extremes, and at a given moment most are heading for one extreme or the other. Where would you like to be headed?

Before we move on, let's say a few words about risk management. The counterpoint to the main point (that loss prevention is a quick way to enhanced profits) is that risk management is not. It can, and in fact often does, cloud the picture. First of all, the name is a misnomer. The average risk manager wouldn't know most risks in their native environment if they fell over them. *Once identified*, most capable risk managers *will* in fact do a good job of transferring or funding that risk. Some money can be saved there, but it is incremental by comparison to that which can be saved by preventing the loss.

What should we call risk managers? It is suggested, with tongue in cheek, that the old name of insurance managers or even risk funding managers is much closer to what most of them do. Of course, there are exceptions. They would be those to whom *both* the loss prevention and risk transfer/funding functions report. Although that is unusual, it would justify the title of risk manager.

Take a look at the program for the next RIMS meeting. It's likely that over ninety percent of the subjects are related to risk funding and transfer and less than ten percent to loss prevention. That suggests that either the funding and transfer are more important or the people attending the meeting have a lot more interest in funding and transfer than in prevention. The latter is the only intelligent conclusion. Further, there will be little or no interest in the loss prevention subject, and those sessions devoted to prevention subjects will be sparsely attended. When is the last time their keynote speaker was a loss prevention person? Look it up. Case closed. By the way, I've attended several of their meetings and even spoken at them.

When thinking of those meetings, the phrase I heard at a human relations seminar several years ago comes to mind: "Those of you who need this the most will, unfortunately, get the least out of it; those of you who need it the least will derive the greatest benefit." I was a little confused by that at first but have found it true. People get better at things because they listen and try to improve. Those who are good at something are good because they have been receptive to instruction on the subject. *So-called* risk managers don't want to know much about loss prevention, so they never will.

AND THERE ARE MANY

We may have overworked the financial benefits of an effective safety process. The hard part is not understanding it but getting others to share your understanding. That will be discussed more later, but for now let's look at the other benefits that should be sold. The complete list is found on the following page.

Benefits of World-Class Safety

1. Loss prevention protects profits.
2. Loss prevention parallels and really *blends with the quality process* and other essential business functions.
3. Loss prevention education can augment and at times *drive the training needs* of an organization.
4. *Auditing* can augment other audit processes and even provide a model. If you can do safety well, everything else should be easy.
5. Good safety process usually coexists with *harmonious labor relations.* Who knows which came first? So, lead with safety and let labor relations follow.
6. An enlightened and comprehensive loss prevention process will enrich, not degrade, the quality of employees' lives.
7. You *reduce the likelihood of corporate catastrophe.*

---

1. Profit Protection. A realist knows that this is the primary driver in any for-profit organization. It is covered extensively in previous sections.

2. Quality Parallels. Interest in quality and the quality process continues to grow. Many of the wise sayings of the quality gurus have been used by safety professionals for years. The quality movement *is* a reinventing of safety.

Several years ago I started to work for a client who pointed out that his company was on a quality binge, and I needed to be sensitive to that issue. He mentioned that most of the company managers had been encouraged to read Philip Crosby's book, *Quality is Free*. I bought the book and read it. I was amazed. It contained no new information, and if I replaced the word *quality* with the word *safety* wherever it appeared, with very few exceptions it worked just fine and could offer a new book just as good as that very popular book. The title would be *Safety is Free*. Actually, safety is *more* than free; it pays dividends.

In the back of Crosby's book he has a lively and provocative section called "Guidelines for Browsers." Let's look at a handful of his snippets of wisdom and read two ways: first using quality and then using safety.

Management has to get right in there and be active when it comes to quality/safety.

Traditional quality control/safety programs are negative and narrow.

The fifth erroneous assumption is that quality/safety originates in the quality/safety department.

Quality/safety improvement has no chance unless the individuals are ready to recognize that improvement is necessary.

If quality/safety isn't ingrained in the organization, it will never happen.

Get the point? These are just five of at least a hundred statements that could be displayed. It's a little eye opening, isn't it? Initially, I thought that quality and safety were first cousins. I now think they are brothers, perhaps twins. Many additional statements in the book don't use the words safety or quality but would be quickly recognized and sworn to by a quality or safety professional.

Here are five of those statements:

Why spend all this time finding and fixing and fighting when you could prevent the incident in the first place?

If you can't produce a dead dragon each week, your license may be revoked.

Changing mind sets is the hardest of management jobs. It is also where the money and opportunity lie.

Attitudes are really what it is all about.

People really like to be measured when the measurement is fair and open.

Where do these comparisons lead in the discussion of safety management paralleling quality management and perhaps many other support functions, like purchasing, human resources, and public affairs? I taught a seminar several times to safety professionals trying to get attendees to find the answer. I did the seminar with Barry Jessee, who had worked with Edwards Deming at Albany International. Barry is a fine presenter; a likable person; very intelligent; well-versed in the basics of quality theory. Nevertheless, I would have to say we failed. If we come to understand how beliefs and values can predict outcomes, can we bridge these gaps? Probably. In one seminar at PetroChem in Trinidad, we even had attendees do maturity grids for safety and quality and examine the comparative maturity of the two processes within their organization. They completed the activity but didn't get our point. The workshops were unable to lead people to be creative and vision ways the parallels could be used to produce synergy.

Why didn't it work? It is unclear, but it may be tied up in not being able to think outside the box. Managers are so ingrained in the Humpty Dumpty School of management and so wed to their discipline that they can't bridge to other disciplines.

How can they bridge the gap? The concepts discussed in this book can help. Managers need to deepen and broaden their thinking and understand how beliefs and values predict outcomes in many different spheres. If values that produce an accident-free environment can be internalized and practiced, what does that do for the quality process in an organization? Or look at it the other way. What effect will keen knowledge and acceptance of quality concepts have on the safety process? Companies know the answers but just can't seem to apply them.

Has anyone done better than Barry and I to marry safety and quality? Perhaps, but I haven't seen it. Most books on the subject, like F. David Pierce's, *Total Quality for Safety and Health Professionals*, tell what the quality movement is all about, review some of the commonalties between the two processes, and encourage safety professionals to learn more about quality. Hopefully, there are similar books for quality people encouraging them to learn more about the safety process, but it is doubtful. Unfortunately, books like David's fail to fully probe the potential and suggest ways to explore and mine the similarities between the quality and safety processes. VIRO can do it.

How? So similar are the two processes that the values that drive loss prevention can drive the quality process with equal power and effectiveness. The safety and quality processes meld as values can be juxtapositioned. Not coincidentally, those same values (really the company-at-large's values) undergird other areas including human resource management, purchasing management, maintenance management, and many others that offer untapped opportunities just waiting to be recognized and harnessed. Obviously, VIRO could open myriad opportunities for synergy.

3. Safety Can Drive Training. At both Merck and Chesebrough-Ponds, the justification for the first training managers, Ken McCullough at Merck and Mel Kuntz at Chesebrough-Ponds, was largely based on safety training needs. Those were and are fine companies. The training function probably started and evolved in most organizations in a similar fashion.

Safety training is varied and sometimes complex. It ranges from simple topics like wearing personal protective equipment (PPE) to the complexities of manufacturing hazardous chemicals without unwanted incident. Gene Bloomwell oversaw the first chemical operator training school at Merck in the Rahway, New Jersey headquarters. Safety was treated separately and woven into virtually every part of the indoctrination. Safety drove the effort. They were convinced that if they did safety well, everything else would fall into place. In retrospect, it seems they were right. I'm not sure the management saw it that way, and I'm sure we didn't do as good a job of selling the role of safety in enhancing quality and manufacturing as we could and probably should have.

4. Audit Enhancement. Next to the measurement of values (as suggested in this book), auditing has the greatest potential, although largely unrealized, to enrich the safety process. Note how it is rated on page 173. It obviously must include behaviors as well as an inventory of physical items. Few safety audits do that, so they take on the look of a glorified safety inspection and, unfortunately, rarely produce much more positive change than those old safety inspections. They have a slightly longer tail but still disappear with only a hint of permanent positive change.

When corporate America thinks of audits, they mostly think of their internal auditors or auditors from one of the big accounting firms who sign the back of their annual report and tell the stockholders that all is well in the money counting department. Due to some bad publicity resulting from an environmental miscue in the middle of this century, Allied Chemical (now Allied-Signal) directors decided that finances were not the only thing they wanted audited. The result was a partnership with Arthur D. Little that produced some of the first environmental audits. The early environmental audits have grown into environmental *safety and health* audits in many organizations including Allied-Signal.

Interestingly, the most complex part of these audits is the safety aspect because there is found the earliest recognition of the human element and how critical it is in the total equation. When IBM decided to utilize their internal auditors to participate in their safety and health audits, they gave them the straight forward regulatory areas to measure. The more complex areas were done by a team of safety and health professionals with a sprinkling of line managers. It is interesting that the people who keep track of that which is most important, the money, can't be trusted to do a good job on any but the simplest safety items. What should this be telling management about the complexity of loss prevention and the potential it has to serve as a leading edge for positive culture change? The human factor is always the most complex element and therefore affords the greatest challenge to measure. If companies can get good at measuring people, the rest is easy. Safety and health can lead the way to better measurement in general and more sophisticated methods to predict outcomes.

5. Safety/Labor Harmony. At Chesebrough-Ponds, when managers got deeply involved in the etiology of soft tissue injuries, they initially found a lot of confusing data. One of the facts that blurred their vision was the number of inconsistent injury reports in different plants in the same state doing similar work. Reported injuries were vastly different. The only significant variable was the labor atmosphere. This is not meant to trivialize the complexity of the analysis, but it came down to whether employees were willing to "play hurt." Everyone had the same afflictions. In operations that had good labor relations, they rarely got reported. In factories where the employees didn't like their management or how they were treated, the labor-management friction served as the catalyst that triggered the injury reporting. Bad labor relations served as a floculent for reporting injuries and optimizing their negative effects.

When corrective measures were imposed, the labor relations improved. And, so did the losses and costs. The employees' perception of their afflictions changed; they developed better work habits; they felt better about management; they felt better about themselves. Everything got better.

It is unclear which drives which, but labor harmony and an accident-free working environment are like conjoined twins. It's rare to see one without the other. The values-driven approach advocated in this book for safety performance enhancement will lubricate and improve labor-management relationships. Every good labor relations professional or safety professional knows this is correct. Why shouldn't safety in the interest of labor harmony be sold to everyone in an organization? Stories like this are sure to exist in virtually every organization. Why keep them a secret?

6. Enhance Quality of Life. The great beauty of the safety movement is that everyone is in favor of it. The great tragedy is that only a handful of people are passionate about it. The difference between mediocrity and excellence is passion. The lack of passion leads to a lack of vision. "Where there is no vision, the people perish" (Proverbs 29:18). Well, the people won't perish without adequate safety vision, but they surely will have more injuries.

How many companies have off-the-job safety programs that equal their on-the-job programs? A very comprehensive survey done by a huge company to benchmark "world-class" safety companies concluded that one of the characteristics needed for "world-class" safety was a vigorous off-the-job safety emphasis process. Some day in the future, wellness and employee assistance programs will be correlated with and intimately related to loss prevention. Managing safety by values will rapidly uncover this truth.

The vision of safety as industrial accident prevention is the view through a narrow width field telescope. Safety professionals need to use the wide angle lens. They need to see that the father with upper extremity soft tissue injury can't play ball with his child; that the woman suffering from occupational lung disease can no longer hike as she once loved; the avid golfer is struggling since damaging his vision at work.

7. Corporate Catastrophe Prevention. Catastrophe prevention doesn't sell well in some industries, but Union Carbide (remember Bophal, India?) and most of the chemical industry can be motivated by it. The Chemical Manufacturers Association may even be a little paranoid about spills, explosions, and toxic gas emissions. Have you seen their new television ads? They reflect the attitude of their members. That attitude is good for everyone.

When I used to call the presidents of the shoe and apparel companies at Chesebrough to make an appointment, they set the time for two weeks later and then usually canceled. When I called Roy Sambrook, Vice Chairman for Stauffer Chemical Company, his answer was always the same: "You can come over now." He was a savvy chemical company executive. He knew that the quickest way for him to lose his job would be to have a serious incident that was widely reported in the media. Not allowing that to happen was his first priority. His behavior is typical of the chemical industry. Safety, particularly process safety, sells easily in that industry, and apprehension about a catastrophe is a built-in driver.

While chemical company catastrophes are easier to envision, other industries are not immune. What is the cost when a food or cosmetic product is adulterated and must be recalled and removed from shelves? How about an automobile recall? These represent catastrophes and result from problems with corporate values.

SELLING SAFETY BENEFITS

Now that some of the significant benefits of a "world-class" safety process are known, that is not nearly enough. They must be sold. When I first interviewed for jobs with my Chemical Engineering degree from Lafayette College, I was only positive of one career that I did not want to pursue: selling was out. To my great surprise, early in my safety career, I came to a sobering realization. I was engaged in selling. And my product was an idea instead of something I could show or demonstrate. Since I liked what I was doing, I set out to learn something about selling and persuasion. I'm glad I did because any success that I have achieved could be more easily traced to my communication and persuasion skills than to technical abilities.

How can safety benefits be sold? The environmental safety and health professional will never be an effective advocate for loss prevention. They are like a noisy bell clanging in the wilderness. They become background noise. Even getting the CEO to issue policy statements and occasional articles in company publications is too contrived and perceived as such by most employees. It's just too easy; it has some utility but lacks power. The real power comes when influential executives—particularly operations officers—speak up for safety. The safety professional may need to choreograph and encourage specific statements and activities.

An overheard conversation at the lunch table when the CEO or COO expresses sincere concern about the safety of some aspect of the business will probably reach a wider audience and have more impact than contrived efforts. Okay, how is that effect achieved? The answer is the same as it is for a lot of hard-to-accomplish, worthy endeavors: with great difficulty and over a long period of time. That being the case, managers can either take the position that they might as well forget it or that they better get started soon since the trip will be long, and they must make it.

Here are some ideas, and you'll need to come up with more. What better to do at safety committee meetings, staff meetings, or even board meetings?

LEADERSHIP MODELING AND EXAMPLE

I liked the movie "Patton" so I read the book. It dramatically changed my understanding of the man. The movie led me to believe that his tirades to inspire his men were impromptu and represented a subconscious gift. The book explained that indeed he saw his primary attribute as the ability to lead and inspire rather than be a master military tactician. Like most people who get very good at something because they like it or have an aptitude for it, he practiced. He rehearsed his "outbreaks" before his staff in his slippers in his office. He did it until he got it right.

I'm not suggesting that executives take safety support acting classes. *But,* they should be aware that if they "sell" safety it will happen, and if they don't, it won't. They should know that sincere and subtle signals will carry more weight than printed policy statements. One of the remarkable qualities of enterprise is how sensitive employees are to the actions of their leaders.

One day Ralph Ward left his fourth floor office and walked down to have a brief, informal visit with me in my office on the first floor. He rarely walked around the building and usually confined his office visits to others in the executive area of the fourth floor of corporate headquarters. His visit changed the way everyone sitting near me viewed me. I even heard they told the story to friends. I accepted my new-found influence but marveled that such a casual, seemingly unplanned visit could mean so much to so many.

Ten years later, I mentioned the story to a friend who gave me another insight. They suggested that the visit may not have been so casual after all. Ward was a brilliant Chairman and CEO. He knew everything the rest of us knew and a lot more. He was the unquestioned leader. Did he decide that afternoon to promote me without changing my title or salary? He did just that. In retrospect, I think he knew exactly what he was doing, as he usually did.

One of our division presidents, Ed Sanford, who headed up our cosmetics company (Prince Matchabelli) had become frustrated with the safety performance of his manufacturing facility but didn't know what to do about it. He confided that to my then-boss, Ted Mullins. Ted suggested he talk to me directly. Ted then told me of his conversation and suggested I call Ed and arrange to visit with him. I believe Ted felt that Ed would be apprehensive about contacting me. I couldn't figure out why but followed instructions. When I made the appointment and saw Ed, I understood. He was a very capable, young, cosmetics company executive. He did not get any safety training in his MBA schooling. He confided in me that he was clueless as to what to do, felt a little overwhelmed, and even subtly indicated that he was a little apprehensive about meeting with me.

I told him that together we would solve this problem. I suggested that I meditate on how to do it in a way in which he would be comfortable and made arrangements to see him a few days later. As I exited his office, the plant manager, who was the recalcitrant, saw me. I made believe I didn't see him. When I got downstairs, I went into Ted's office and reported on the meeting and the chance almost-encounter. I wondered whether it was a plus or a minus. He suggested that it was probably very good. As usual, Ted was on target.

I revisited Ed Sanford and made a single suggestion to him. I brought an accident report form and showed it to him. I simply suggested that on every visit he made to the plant, he ask to see the reports for all the accidents that had occurred since his last visit. He should read them casually and ask questions about any incomplete sections or anything he didn't understand. That was it. He felt very comfortable doing that.

During the next six months, the record at that facility improved dramatically. Within a few years, they won the Chairman's Corporate Safety Award and got the CEO to visit in the corporate jet, the bowl with their accomplishment inscribed, and gifts for all the employees. I don't know exactly what did it, but subtle executive influence had to play a part.

Getting Leadership to Sell to Each Other

How does one make these kinds of things happen routinely? There is no exact answer but they do happen in companies that achieve excellence in safety. The voice of the safety professional gets lost in the crowd, with a few exceptions. The trick is to facilitate, but stay in the background. The day I started at Chesebrough-Ponds, Ted Mullins asked for what I would like to be known when I was done there. I said that I'd like to have the most accident-free environment in the world and for no one to know who the safety director was. There isn't much that I said or did or wore twenty years ago with which I am still comfortable, but I am with that answer.

An example of how I tried to do this comes to mind. We had a corporate safety committee made up of the vice presidents of manufacturing of all the business units. It was the only time they all got together as they were in very different businesses. Apparently, the only common thread was safety. The manufacturing vice presidents thought the meetings were to set policy. I knew differently. The meetings provided a rare opportunity for me to train them. I soon realized two things: 1) I would probably never convince all of them of anything; and 2) they didn't exactly see me as a font of knowledge on any subject, including safety.

I decided to let them train each other. I would pick individuals receptive to a given subject and help them experience success and enrichment in that area, and then have them tell the group about their experiences and accomplishments. It worked like magic. The presenter felt good. The audience listened. And then everyone wanted *one of those* too (whatever it was I was selling, i.e., a hearing conservation effort or better new employee training). We propagated Job Safety Analysis, hearing conservation, ergonomics, and a host of other programs throughout the company using this technique.

PEANUTS   By Charles M. Schulz

PEANUTS reprinted by permission of United Feature Syndicate, Inc.

A SUCCESS STORY

In addition to the businesses I have mentioned earlier, we had a hospital products division at Chesebrough-Ponds. One aspect of that business was making the now mostly obsolete mercury in glass thermometers. We sold them under many different names, like Ballo and Fachney as a result of numerous acquisitions. We made the thermometers and a few related products in Watertown, New York. This was a place noted for demonstrating to the U.S. Army a climate more suitable for Arctic training than Nome, Alaska, where it was being done and I suppose still is (reported in the New York Times *about fifteen years ago).*

One day, Mike Griffis, who was our industrial hygienist, came into my office and showed me a translation of a Russian article on the toxicity of mercury that was several years old. Don't ask me what he was doing reading old Russian articles on mercury. Perhaps it's obvious to those of you who are industrial hygienists.

*The study suggested that mercury could pass through the placenta and hence **could** be responsible for **some** birth defects, although there was little or no supporting literature. I brought our discovery to the attention of my leaders and they asked me what I wanted to do. I presented the information with a balanced perspective, but for one of the few times in my career, did not want to make a firm recommendation or take a rigid stand. I suggested we get the appropriate decision makers in the company together, present all sides of the case, and allow the leaders to make the decision.*

My managers agreed and set up a meeting in our board room with several officers, including the leader of my company sector, Bob Bennett, a Vice President who reported to the CEO, Ralph Ward. Others involved were the Hospital Products company president, John White, his leader, a group vice president, and some other officers (in a company that only had fifteen or so officers) and several other interested parties.

As background, we were following virtually all the requirements in the mercury standards and achieving generally acceptable results. Our exposure levels were below the Threshold Limit Values, but in a few areas above the Action Levels for short periods of time. We did routine sampling, urine testing, and some blood testing. For those unfamiliar with mercury, it is unlike lead in that it does not accumulate in the body, so high urine tests are not all bad. They do show a probable recent high exposure (the bad news) but also demonstrate that the body is doing a good job of shedding it (the good news). Invariably, we found that when people had high levels it was due to personal hygiene problems. When we forced employees to shower and change clothes at work, the problem went away. You see, a little mercury under your finger nails or in your shirt or blouse near the breathing zone goes a long way.

The plant was relatively new, but part of it had been moved from an old mill building within the last decade. Some of the employees had even worked in that old building. Some of the furnishings and work stations had been moved from the old facility. Mercury tends to collect in cracks and crevasses and evaporate slowly. A clean, modern building with new surfaces is easy to clean; an old mill building almost impossible. Modern furniture and tables with hard, smooth surfaces are ideal; old wooden work stations are inadequate. We had the modern building with adequate but not exceptional ventilation, and we still had some old work stations.

We suggested that the company was in compliance but, did at times, approach levels that would concern some people. We suggested three possible approaches in response to the concerning but fragmented and essentially unsupported literature on mercury. By the way, we did some crude epidemiology work and found no reason to be obviously concerned with birth defects.

Our **first** approach would be to do nothing but monitor the situation.

The **second** would be to make some token efforts to get the biggest bang for the buck by replacing some furniture at work stations and enhancing our clean-up procedures and providing better education and motivation for personal hygiene.

The **third** option was the do-it-all option. In addition to what would be done in the second option, we would provide double lockers and more showers; wash work clothes daily; install enhanced and independent ventilation systems by segmenting the plant, adding heating, and reducing recirculation (remember the Arctic climate); and finally, replace some floors with better surfaces. These were obviously the big ticket items and would cost well over a million dollars but would almost guarantee cutting exposures in half.

At the end of the several hour meeting, Bob Bennett asked all but the officers to excuse themselves, and as I walked out, he said he would call me when a decision had been reached. Bob is a compulsively honest person whom I greatly admire. As further background, the hospital products business was not a big money maker for Chesebrough. In particular, the thermometer business was already getting stiff competition from off-shore manufacturers and could see electronic temperature taking clearly on the horizon. This division had a hard time getting capital for anything.

In less than an hour, Bob called me and asked me to come down to his office. I prepared myself for a let-down. Being a man who usually got right to the point, he said, "We have *decided to follow all the recommendations you made." I was pretty sure he meant the practical, cost effective second option as we had only offered the third as an illustration of just how far we could go. I inferred that was my interpretation of what he said and he clarified what I had apparently misunderstood; over one million dollars would be spent to substantially update our facility.*

Just a few days later, I was walking down Greenwich Avenue to the train station, which was about a mile from our office. I was headed into New York City for some meetings and was absorbed with thoughts concerning the day ahead. A car stopped and offered me a ride. It was John White, whom I hadn't seen since the meeting. As I entered his car, I was a little apprehensive. I would have preferred not to have talked to him as I know the money spent would probably come out of marketing or production enhancements that his company badly needed. I apologized for raising an issue that had resulted in complicating his job. He responded that he always felt that what you didn't know could hurt you and I had done my job and he appreciated that. John was not only an effective executive but a fine gentleman.

In reflection, how does this story sum up the message on selling the many benefits of safety?

- *If you are credible and sensitive to the full range of business concerns*, healthy and sound organizations with moral leadership will respond very positively. They will place a high value on environmental safety and health expertise.

- Costs will always be a concern, but doing what is right will often supersede costs.

- This decision was not an isolated event in the history of our company or my relationship with the decision makers. They viewed me and the people who worked with me as part of their solutions and not as unnecessary overhead. That was the background for these scenarios. Without it, a different kind of decision could have easily been reached. We had made numerous deposits, so when the time came for a withdrawal, we could draw against the balance.

- The old saying that you can't soar with eagles if you work with turkeys comes to mind here. As a consultant, I have worked for some turkeys. I was of little use to them. They wasted their money on me; I wasted my time on them. I'll be more careful in the future. At Factory Mutual, Merck & Company, and Chesebrough-Ponds, I soared with many eagles. They made me look good.

- I constantly sold benefits and, consequently, never lost an employee during cut-backs and always fared better during budget paring exercises than my colleagues in other departments. I invested meaningful time educating our management about the savings that a sound safety process yielded. It came to color all their thinking.

12

NEVER SETTLE FOR SECOND BEST

"Success comes in cans, failures in can'ts."

It was 1975; I was being recruited to be the first safety director for Chesebrough-Ponds. Ted Mullins and the others with whom I had interviewed had decided that they would offer me a job. I had decided that I should probably make the move in spite of the risks involved and my comfort in my present situation at Merck. As I tried to envision what it would take for me to be successful, I concluded that the value the organization placed on what I was doing would predict the outcome as much, if not more, than the skills and energy that I brought to the job. How to evaluate that; that was the question.

I needed a simple and quick test. I knew Ted had six positions reporting to him. I had met several of his direct reports. I told him that it was not important how his salary offer to me compared to what the other people who reported *to him made. But their job ratings were important to me. I inquired as to how the jobs were rated. He responded that three of them were the same as mine; one lower; one slightly higher.*

That was a significant factor in my decision to accept the position. It told me that the organization valued the safety position as highly as it did labor relations, employment and placement, and the other corporate industrial relations functions. Throughout my thirteen-year career at Chesebrough-Ponds, the view management had of what I did had everything to do with my success and that of those who labored with me.

Sure, I had something to do with their perceptions, as did the performance of the people with whom I surrounded myself. The primary criterion for people I hired was that they had to be brighter than I was. We were in an atmosphere that, right from the beginning, didn't relegate what we did to second place.

Before I go any further, I need to address the "second best" in the chapter title. It is *not* meant to infer that loss prevention should be first. On the contrary, I find that organizations who hang posters suggesting "safety first" are suspect. It means that they are out of touch with reality, trying hard to convince someone of something they aren't buying, or just subscribing to posters they don't read or understand. Very few companies are in loss prevention as their primary business. As our old friend, Tom Peters, has reminded us so many times, you need to know what business you are in or you will soon be out of business.

Environmental safety and health needs to be *one* of an organization's priorities; it needs to be something they understand, believe in, and on which they consequently place value. The personnel are one of the best indicators of how management values a subject. For instance, Chesebrough-Ponds was a marketing company. They always hired the top marketing people in the country. They knew what our vital interests were. Merck was a company that depended on new discoveries. They lived or died on the basis of the productivity of their research, not their plants. They hired, pampered, and rewarded the best pharmaceutical researchers in the world. And their forté was inventing new drugs. Where their heart lay, there was their treasure. While loss prevention is rarely a company's primary function, it should be viewed as vital to accomplishing the primary mission.

THE HISTORY OF SAFETY STATURE

Until the middle of the century, the safety position was usually filled by a person for whom the management couldn't find another job. Nobody was quite sure what the position did, but they were pretty sure it wasn't critical to their success. Not until a couple of decades ago did universities devise educational programs for safety practitioners and start graduating significant numbers of highly motivated and intelligent young people. Still, it was not then and is not now clear exactly what they do and how they can benefit an organization. That is due to:

The Regulatory Orientation

Just as the field started to coalesce and find itself, a mixed blessing occurred. The Occupational Safety and Health Act of 1970 was passed and began to dominate the loss prevention process. That was good in that it got attention where it may have taken decades to wake up some people. For those who were already ahead of the OSHA process, like Merck, it set them back years. They had to concentrate on documenting what they had already done; do things that had nothing to do with loss prevention but were part of some obscure consensus standard; or follow some form that was not as good as what they already had but was required for compliance. Slowly, very slowly, the regulatory machine has refined itself. But it still inhibits as many people *or more* than it stimulates.

Leadership is still confused by regulation and how it *really* relates to their primary business. Restrictive and onerous laws encourage management to see environmental safety and health as an obstacle they have to overcome rather than a catalyst for their business and a potential area to gain a competitive edge.

The Linkage with Insurance

From the early days when most of the industrial safety managers were trained by insurance companies, the insurance industries have been linked to safety. Management is much more familiar with their finances and insurance than they are with loss prevention. That is explained in more detail in several other chapters. Initially, the link with insurance probably benefited the safety movement and the individuals who spawned it. Today, it is a millstone tied around the neck of loss prevention practitioners, and it gets more onerous every day. It inhibits progress, stigmatizes the profession, atrophies self-image, and depreciates the value of the contributions that the profession can make to an enterprise and society. The most important influence on the stature of safety in an organization is the management.

The Management

These folks, with the exception of the most enlightened, are clueless as to what a safety practitioner does. If they do know, they don't know much. They almost uniformly fail to recognize the potential the function has. They restrict it in untold ways by determining whom the safety practitioner reports to and how the safety practitioner is positioned in the organization. This would be like trading for Steve Young and not letting him run or throw the ball. Sure, he could probably play cornerback. But is that the best place to play him? Similarly, often management does not know where to play the safety practitioner.

You rarely get more from someone than is expected of them. In an organization, the expectations are based on what people are paid and how they are officed, the size of their budget, and the resources given them. If the safety practitioner isn't paid well, given adequate resources, or is buried in the organization, what would lead a manager to think they'll be able to enrich the organization or become a great profit center for them?

HOW *SHOULD* THE FUNCTION BE VIEWED?

To start with, the function should not be linked to lawyers (regulatory), risk managers (insurance), human resources (training, etc.) or to any other restrictive function. This suggestion has nothing to do with reporting relationships. That has already been covered in Chapter 9. Now, how should management view the conservation of their physical and human resources and the people who will be their strike force in that area?

I always took a very broad view of my function and could be found in almost any area of our company, showing up at meetings of almost any group. All I wanted to be able to do was to demonstrate that they were better off for my having been there and that I probably couldn't have done the company more good by being anyplace else (prioritizing my contact time).

WHAT DO SAFETY PEOPLE DO?

Enrich Resources

Would you want a tour guide to sit in the back of the bus without a microphone? Because of the nature of the function, the loss prevention professional can and should be involved in every part of an organization and every facet of its life. At the function's best—which is where it should be—the well-qualified practitioner can be totally upbeat and even inspirational. Such a "tour guide" is not driving the bus, and they probably didn't specify the route, although were hopefully consulted. But what would the tour be like without them?

The overachieving safety professional who is visionary can:

- Serve as a conscience for the organization, in all facets of its life.
- Promote and help refine wellness and employee assistance efforts. *(A forest products company developed their wellness and employee assistance initiatives as an outgrowth of reducing workers' compensation costs and instituting improved claims handling.)*
- Assist in the practical and efficient design of facilities and products. *(At a toy company, a safety engineer helped design new fasteners that not only reduced cumulative trauma injuries by assemblers but reduced production costs and made extraction of the toy from the package more user-friendly for small hands.)*
- Assist in controlling procurement costs. *(At a chemical company, they saved millions of dollars on personal protective equipment as a result of efforts driven by the safety department.)*

- Lead the way in defining and developing professional competencies. *(Bill Driscoll did this at OxyChem as mentioned earlier.)*
- Be a pioneer and later team member in training efforts. *(I did this at two major companies.)*
- Assist in various public affairs efforts. *("Right-to-Know" programs don't have to be the end. They can just be the beginning.)*

Assure Profits

Profits come not only from manufacturing and selling things, they also come from avoiding overhead associated with the research, manufacturing, selling, and marketing (i.e., with the entire process). Developing strategies to be the most efficient (that includes avoiding unnecessary costs/losses) is directly related to minimizing overhead. If they are good, safety practitioners have the potential to provide the best return on investment of anyone in the organization—if they're not, no organization can/ should afford them. This is discussed in some detail in the last chapter.

HOW DO YOU GET TO WHERE YOU WANT TO BE?

First, create awareness and expectations. Tell people what should be expected of a broad and deep approach to conserving physical and human resources.

Second, set up models and experience success. Map out what should be done. Decide what you believe in and on what you are going to place a high value. Then set up models or pilot projects to accomplish your goals. Don't under resource them. Nurture them.

Third, commit yourself to excellence. Seek to achieve "world-class"; it is worth it. Others have done it and you can, too. Even the best are not that good. You can do better if you *really* want to.

Fourth, don't be satisfied with anything less than achieving your fondest expectations. This is a largely untapped resource. The benefits are hard to even imagine. Think big.

SUMMARY

No one associated with the conservation of the physical and human resources of an organization should be considered anything less than one of its most valuable resources. Would you let someone vandalize your home while you watched? Would you stand by and watch someone as they stole your new car? If these risks presented themselves, what value would you place on preventing them from happening? That is how you should view your loss prevention efforts. Nothing else but full effort makes any sense.

13

BE GUIDED BY LOGIC, NOT EMOTION

"Logic is necessary, since without its assistance you cannot so much know whether it is necessary or not."
-Epictetus

Men and women bring emotion in varying degrees to their every activity. Like stress or friction, emotion is essential to today's world. It is at times a positive force and at other times a negative one. This text will define good/positive emotions and bad/negative ones.

Good emotions are sensitivity, compassion, empathy, love, and charity. Women are generally believed to have these in larger measure than men.

Bad emotions are anger, fear, agitation, distrust, self-interest, contentiousness, and irrationality. Men seem to exhibit these more often than women. Who has started and fought the most wars?

Emotion is a difficult (emotional?) topic to tackle, so some readers may raise eyebrows and disagree. This book's goal is to stimulate thought. Hopefully, you will have some fun, too. The good and bad emotions *are not* black and white. If misapplied or allowed to dominate a situation, the emotions labeled as "good" can lead to confusion and bad decisions. On the flip side, some of the "bad" emotions could be essential for protection or to stimulate action.

THE DEATH OF COMMON SENSE

Common sense is closely related to logic and reason. At one point, my editor became rebellious about my equating common sense with logic and reason. Everyone has seen some rather strange behaviors based on "common sense" that they thought were nonsense and not held in common by any associates. Nevertheless, I need to use the term and believe it can be equated with logic and reason if I define it for use in this chapter. The dictionary says sense is "a capacity to appreciate or understand something sound or reasonable," "grasp, understand/detect automatically," and "faculty of perceiving." If the word "common" is placed in front of it, those definitions are simply emphasized by suggesting that most people would agree or hold a perspective in common. This is one of those terms that has been so misapplied that it doesn't mean much to many people anymore and often elicits cynicism.

The reason this book is dealing with the need for common sense in the safety movement is because it is mostly absent yet is essential to achieving good outcomes. Witness the popularity of Philip K. Howard's non-fiction best seller, *The Death of Common Sense*. It is interesting that many of his examples are safety related and that he often comes up with the wrong solutions. As a matter of fact, his general conclusions are either wrong, misrepresented, or perhaps misunderstood. Since the book is so relevant, current, and indeed popular, this chapter will briefly review it and its applicability for safety professionals and others who seek to optimize the incidence of desired outcomes in their lives.

There is much wisdom in the book and some bittersweet humor; there are also some mistakes. As with most mistakes, we can learn from them. Howard starts out by relating some stories; this is an effective technique. He describes a horror story of Mother Teresa being thwarted in efforts to provide housing for the homeless by crazy building codes. Then about AMOCO spending $31 million to satisfy an EPA regulation that missed the pollution; a perfect bureaucratic failure that maximized the cost to the regulated and minimized the benefit to the public. Then he offers a safety example and makes his first mistake, one that can at least be partially excused because he was not entirely wrong. His mistake was one that even safety professionals could and have made, but nevertheless a mistake. It is possible to forgive him but not the company safety officer he writes about.

Howard quotes a regulatory compliance manager as disenchanted with OSHA's obsession with paper and physical conditions (he was correct here). Then he suggests the solution is to give tool sets to employees for injury-free work and tax-free cash awards in a lottery tied to company injury records. Unfortunately, Philip, that's not the answer, either.

Howard moves on to describe a Massachusetts lead paint program that forces removal or abatement. Horribly, some children's blood-lead levels worsened after abatement began. Sound like asbestos? Just one more example of insensitive and rigid legal dictates precluding the exercise of judgment (logic) at the time and place of the activity.

In another example, the author contrasts our Constitution to the longer EPA benzene standard. He champions flexible law that changes with the times and focuses on what a reasonable person would do rather than loads of paper and detail. By its very nature, VIRO accomplishes this and sharply contrasts approaches that rely on loads of manuals and procedures, yet don't work. *I had a group of fifteen clients in the same business at one point. The one with the best records of performance had the least documentation and one of the worst performers had the most documentation. The good performer empowered and lived loss prevention; the weak performer wrote about it but didn't do much about it.*

The most disturbing thrust in the book is Philip Howard's almost constant deriding of process. He rightly suggests, "Process has become an end in itself." But, he fails to distinguish between good and bad process, inferring that all process is bad. He correctly points out that managers often find themselves arguing about not right or wrong, but whether it was done the right way. He states, "Setting priorities is difficult in modern government because process has no sense of priorities." That may be right if process is set in concrete or not based on solid values, or worse yet on the wrong values. But, if process is flexible and based on values that predict success; how can it get any better? He quotes Plato's argument that "good people do not need laws to tell them to act responsibly, while bad people will always find a way around law." Ah, there is a pearl. Responsible, reasonable activity flows naturally from people with the right values.

Throughout the book, several quotes include the word *values*, as in those below:

"Like printing money, handing out rights to special interest groups for thirty years has diminished not only the civil rights movement but the *values* on which it was founded."

"But we will never clear away the procedural fog until we redefine our *values*."

"We aspired to decent *values* of hard work and community." [This referred to the 1950s America after two decades struggling to overcome the Depression—possibly the best time in the history of our nation.]

"As Tocqueville warned, however, the momentum was likely to carry it to a point where the *values* that prompted reform were themselves eroded." [Here the context is response to racism, sexism, unenlightened treatment of the mentally ill, environmental problems, and a host of other social ills.]

Needless to say, the book is littered with references to the need for common sense and good judgment in place of senseless laws and rules. Unfortunately, it concludes: "Relying on ourselves is not, after all, a new ideology. It's just common sense." This is great advice, but there is no implementation strategy or plan. A values-driven strategy, like VIRO, seems like the logical implementation approach. As indicated in Plato's argument, volumes of laws and rules become unnecessary in a group of people guided by logic, not emotion.

THE PITFALLS OF EMOTION

Here are some stories to illustrate the mob-effects of emotion on reason and set the stage for the conclusions and suggestions.

The first is a baseball joke.

It takes place in a small minor league baseball town. The main characters are a third baseman and a hard core fan. The player has just been advanced from a rookie league where he had a very good year. The fan has not missed a game in twenty years and is emotionally attached to the team in a pathological way. The fan takes an instant dislike to the new player, who has a pretty good year hitting clean-up but certainly makes some bad plays, strikes-out with men on base, and even has some very bad days. The fan never lets him forget his errors. He constantly shouts and reminds the player of each miscue or strike-out for days after the occurrence.

The last game of the year rolls around and the player has had a good enough year so that he can expect to be advanced and leave this small town and his fan/detractor. He walks over to the fence during warm-ups and motions to the fan to approach him. The fan does so somewhat reluctantly, always making sure he has room and time to escape. The player says he recognizes the right of the fan to criticize his play and suggests that perhaps it has even strengthened him as a player. But, he has one request of the fan. He has arranged with the manager for the fan to play one inning at the end of the game, and he would like the fan to literally spend one inning in his shoes. The fan agrees, pointing out that he played third base himself, was quite good, and would welcome this opportunity. Final arrangements are made with the manager and clubhouse man. The player begins to question the wisdom of his idea.

The last inning arrives and the fan takes the field. The first batter hits a "grass cutter" to third, but the pot belly of the fan inhibits him from getting down; the ball passes untouched, right between his legs. The second batter hits a chopper that our fan fields chest high and fires to second base, aiming for the double play. But the throw flies about ten feet over the second baseman's head into the right field corner. The right fielder makes a spectacular recovery and perfect throw to third to get the batter as the lead runner scores. Our friend fields the throw in plenty of time and lurches at the runner, who slides around him, avoiding the tag. The next batter hits a high pop-up in the middle of the infield. Our dauntless fan *calls everyone off and dives for the ball at the last minute, never touching it.*

The player is thinking, "I can't believe how well this has worked out. I couldn't have scripted this better myself." As the inning mercifully ends, the fan approaches the dugout with a smile. The player blurts out, "I guess that proves my point." The fan/third baseman responds, "On the contrary, it confirms what I've been saying all year." Our confused player asks, "How do you figure that?" The fan closes the case and the season with: "You messed up the position so badly that even I couldn't play it."

That story almost always gets a laugh. People laugh because it is so illogical that there is virtually no response. The fan's emotional involvement has destroyed his ability to reason. Everyone sees a lot of that in their lives and either don't recognize it because of how it is dressed-up or fail to respond because they see the situation as hopeless or beyond their sphere of influence.

Let's take a look at some real life examples. In a recent article in *Professional Safety*, two well-qualified authors write about exercise and cumulative trauma disorders (CTD). They cite numerous studies and conclude that "The Jury is Still Out" concerning the efficacy of exercise in reducing CTDs. I have spent a lot of time on this subject and I know the jury is in.

A paper like the one in *Professional Safety* confuses rather than edifies. It is illogical. Good science should converge with *true common sense*, just as sound science should ultimately not conflict with true religion. Here's one perspective on the subject. Why do athletic teams stretch before playing and when stress has built up? Why do most professional and college teams now have strength coaches? After injury occurs, why is exercise and "work hardening" used before returning a high priced athlete to regular play? Most physical work is para-athletic. The motions and stresses are identical. So is the need for exercise.

So, if properly applied, exercise is clearly the primary preventive strategy, along with job redesign and many other tactics that should be employed concurrently. Exercise should include a warm-up, stress relief, and strength building; which are different exercises done in different ways at different times. Anyone who has ever done any physical activity and used various forms of exercise knows it works.

What has happened here? We have over-intellectualized a subject and relied on unsound science, leading to an illogical conclusion that will not reduce CTD injuries. But, it probably will stall some efforts that could have led to better outcomes. How sad. Everyone loses.

How about the asbestos scare? It is still going on. School districts with shrinking resources and growing needs are spending millions on asbestos removal. There are probably a handful of cases where it could be justified, but not many. How many students have died of mesothelioma the unique form of cancer that asbestos can trigger? It's a sure bet that no one knows of one. To start with, the exposure only really exists if the material is airborne and the particle size is less than five microns. That is too small to see. Sure, loads of people who worked in factories that produced asbestos-laden products suffer from fibrotic lungs, emphysema, and (a few) mesothelioma. Pipe fitters in chemical plants who cut asbestos laden pipe insulation for years without respiratory protection frequently have asbestos related afflictions. But not students in schools with asbestos-containing floor tiles or ceiling tiles, non-friable pipe coverings, or even sprayed-on insulation in out-of-the way places like boiler rooms; they have no exposure. The exposure occurs when it is removed and becomes airborne. This situation not only solves a problem that didn't exist but creates one in the process and uses precious resources to do it.

Then, to top it off, legislation dictates that expensive insurance must be purchased by the contractor and passed on to the school district. The insurance covers a period of three to five years, even though effects from asbestos exposure are latent and do not show up for five to twenty-five years. There is almost no possibility of any claims ever being made. How would you like to be the underwriter on this one?

How does this happen? The only answer is mass hysteria seasoned by greed and ignorance. The asbestos story is not new to safety professionals. Everyone knows about it. Why didn't safety professionals speak up? And if they did, why didn't anyone listen? The asbestos debacle represents an advanced form of emotional dysfunction. Is this so unusual?

How about the silicone implants? Dow Corning provided a valuable product to 750,000 women in the United States and 1.3 million worldwide. A handful of medical problems resulted in a feeding frenzy by trial lawyers who filed a class action suit. Advertisements appeared in newspapers all around the country inviting women who had silicone implants to join the giant lawsuit. Most of the women had no symptoms but were frightened into thinking health problems were on the horizon. Then the lawyers misled them into believing they would receive large payouts from Dow Corning.

In their greed and the emotional rush to judgment, the trial lawyers neglected to ascertain whether silicone implants *really* posed a health threat to women. They relied upon tearful women offering anecdotal testimony. That was backed up with highly "suspect" scientific evidence.

But the most authoritative research has found *no link* between implants and the various maladies that the trial lawyers attributed to them. A study conducted by the Mayo Clinic, published last year in the *New England Journal of Medicine*, found *no* definable disease related to implants.

I'd like to reflect on chronic fatigue syndrome. Almost a decade ago, when I was serving as an LDS bishop, I had several female members of my congregation who developed mysterious symptoms. They didn't fit any standard disease but the symptoms were almost totally debilitating: weakness, joint soreness, listlessness, and various other problems like nausea and disorientation. Their doctors were telling all of them essentially the same thing: "We can't find anything wrong with you, so it is all in your mind." I knew these women before, during, and after. They were among the most hard working, resourceful people I had ever met. I knew that it wasn't strictly a psychological problem. I was only seeing a small sample and I'm not a physician. But, I told my wife, "It won't be long before we find out this is happening all over the country and someone comes up with a name for it." Several years later, enter chronic fatigue syndrome. I know I wasn't the only one to know something was very wrong. Although the diagnosis of chronic fatigue syndrome is controversial, it is no longer simply being dismissed, and medical research is attempting to better characterize it. Why did it take so long to get to where we are? Emotion superseded logic, reason, and good science.

Remember "Desert Storm", or the Persian Gulf War? Many veterans complained about unusual symptoms. Some very competent doctors and psychologists talked about this on a casual basis. They all doubted any correlation with unknown or unusual exposures. Now what is happening? Some television news magazines and the First Lady have sparked a study that surely will reveal facts that have, up to this point, evaded the public. Maybe they will find a root cause and develop some solutions. Why did it have to take so long?

At this time, the U.S. Navy is experiencing a twenty-four hour "stand down" for over 400,000 service people; it was called by Admiral Mike Boorda, Chief of Naval Operations. They are going to give instruction on proper behavior and discipline. Isn't it coincidental that today's newspapers are featuring a story about a drunk Navy cook who sexually assaulted a female sailor aboard a commercial jet and that the twenty or so other sailors on the flight chose not to intervene. Didn't they learn anything from "Tailhook"? Do you think they are asking themselves, "Am I acting or reacting?" If they did, what do you think the answer would be? Does this trouble you at all with regard to the way they handle loss prevention?

I'd like to reflect on my lawyer friend, John Lambert in Portland, Maine, and complete the story that I related in Chapter 9. I didn't have the authority to hire John, and I did have to "apologize." I got back to corporate headquarters and expressed to our risk manager my desire to change law firms. He controlled the broker, fronting company, third party administrators and all those associated with any and all of them, including the law firm that was paid by claims handlers. He did just what any good corporate executive would do: he called a meeting.

His boss, the treasurer, was there along with representatives of the broker and insurance company. Several people traveled from a distance and, due to scheduling problems, the meeting took several weeks to coordinate. Just for effect, we held it in the board room. I was asked to explain why I had caused this meeting and was rocking the boat.

As background, I was spending about half my time in Maine and the CEO was suggesting things like my getting an absentee ballot in Connecticut or just registering to vote in Maine. None of the other parties at the meeting had (to my knowledge) ever been to Maine; met any of the lawyers we were talking about; been in one of our factories; or attended a workers' compensation hearing there or anywhere else.

I stated my case and asked for the change. There was some consternation and confusion, and I was told that it probably couldn't be done because it would be considered impolite. There were protocols associated with dealing with law firms, particularly when we were insulated from them through brokers, insurance companies, and claims handlers. (Note that all the above were effectively represented at the meeting.)

*After some pleading on my part, they indicated that all old cases were off the table (that is where all our money was tied up in reserves) but that they would **consider** trying John on **some** new cases.*

We were about an hour into the meeting (on which I suspected they planned to waste a full day) when I mumbled something about having some other things I needed to do and quickly packed up and left. I didn't know exactly what to expect but couldn't think of anything better to do at that exact time. Well, I got a response.

The next morning, my boss, Ed Hughes, the Corporate Vice President for Manufacturing called me. He said the treasurer was in his office telling him a story about a substantial breach of corporate etiquette on my part. Several officers of companies had visited us from afar, at my request, and I had walked out on the meeting. He wanted me to come right to his office and explain myself.

My explanation was short and simple. I was told that I couldn't have what I said we needed, so I thought the meeting was over and left. I apologized if I had offended anyone. The CEO got word of the altercation and he was apparently more interested in saving millions of dollars than observing strange etiquette. The change was made; my tenuous relationship with the risk manager was not enhanced; and the treasurer, in spite of my apology, didn't smile at me anymore when we passed in the hall. His greetings, when offered, were grunted.

I had not burned all my bridges—I took care of those left a few months later. I'll describe how I did that in the next chapter on empowering others.

What happened here? The relationship between brokers and clients is often lubricated with incentives that vary in size and nature but are designed to encourage loyalty and dependency. Similar bonding mechanisms probably exist between law firms and insurance companies. *Logic was not in play here. Greed and emotion were.* I was tilting at wind mills, but I had a friend with a howitzer. Remember the "bullet theory." Don't count on this kind of thing happening very often, no matter how much sense it makes...*unless an organization has clearly stated safety values that change the culture and behaviors*.

Another example of emotion precluding logic occurred back in 1979-80, when the American Society of Safety Engineers commissioned a membership survey that revealed many things of concern to the members, including involvement in standards making, government relations, and a host of other subjects. I chaired a committee to study the report and their concerns. We were to make recommendations to the Board of Directors. My committee was made up of the "cream of the crop" in our profession. They did a wonderful job, and we came up with about twelve suggested changes. In retrospect, I think this period marked one of the turning points for the society.

We brought the changes to the Board. As a reflection of the confidence they had in us, they approved action on all items, including numerous new directions and expenditures of time, effort, and new monies. A few items needed to go to the Assembly.

One was a revisit of our name and acronym (ASSE), with an eye toward something more descriptive and lofty. Remember, only a very small fraction of the members of the society are actually engineers of any kind, much less safety engineers.

The name change proposal had to go to the Assembly which had representation from all one hundred-plus chapters. The proposal was not that we change the name, nor did we make any specific recommendations other than suggest some options. We only wished the members to consider the subject via some type of referendum. Due to fears that non-engineers may lose something in a name change—some nostalgia and a few states where there was special significance to being called an engineer— resistance developed. At the Assembly meeting in Houston, something very interesting happened. *Emotion ran rampant.* There was actual subtle physical pressure that could be observed from the stand. There were more members from Texas there than any other area. They had very special feelings about the name.

When time came to a vote, the proposal (to allow the members at large to consider the *possibility* of a name change) was voted down almost unanimously. With a single exception that I could see, everyone, including committee members and Board members, went along with the tidal wave. I'll always remember Charlie Dancer's voice and vote. It told me a lot about him and how deep character and conviction go. The majority's decision made no sense at all. Emotional fear and anger resulted in a decision to play the ostrich and refuse to even talk about a very important subject. I decided that I would not be a one issue leader, quickly put it behind me, and have only mentioned it a few times in jest since then.

At this point in time, the world is mourning the death of Yitzhak Rabin. Could anyone think of anything more senseless in this troubled and contentious world than the killing of this internationally respected man of peace? His wife, Leah, has suggested that she feels that inflammatory rhetoric by some of his political adversaries (including Rabbis) was an ingredient in what happened. Their talk likened him to Hitler as a destroyer of Jewish people. How absurd. Yet some demented people may and apparently did take that as a signal. Talk can hurt—believe it, because it is true. Inflammatory language can distort emotions and thereby impair reason. The words can affect outcomes and those words flow from individual beliefs.

The U.S. continues to do studies to determine if violence and sexually explicit shows on television and in the movies affect the beliefs, values, and behaviors of our children. It makes a lot of sense to try to determine *how* it does, but it is difficult to understand the brand of logic/common sense that ignores movie and television influence on youth. Does it make any sense to you? How do you feel after being stimulated by some type of media? Have you ever observed a youngster after an exposure to media violence? I have some trouble remembering my son, but I've observed my grandsons. Anyone who seriously thinks this is a real question must have somehow missed the last twenty or thirty years. Yet it won't go away. Could financial concerns and twisted values be making this an emotional issue?

Obviously, some of these are extreme examples; others could be viewed as isolated incidents; one or two are funny...but, they all form a backdrop for the discussion of why this value—be guided by logic, not emotion—is important and will be closely associated with the best set of safety outcomes.

Moving back specifically to safety, this book will provide a perspective on a few "programs" that often feed off misdirected emotion and pump fog into the loss prevention process. Then it will suggest what should be done to eliminate the fog.

FOG GENERATING PROGRAMS

The following demonstrates several fog generating programs.

Fog Generating Programs

- Traditional Accident Investigations
- Accident Rate Goals
- Discipline Programs for Non-Compliance
- Workers' Compensation Cost Control
- One Stop Shopping Safety Solutions
- Incentive Programs

Traditional Accident Investigations

Very few accident investigation programs produce consistently accurate results. The fog starts with the forms that are used and concludes with the summaries that usually require simplistic coding. The forms often don't ask the correct questions, and management rarely demands accurate answers that consistently reveal root causes. Fear of reprisal (i.e., emotion) colors many investigations. Excellent training of investigators is rare indeed. Pressure to finish the investigation is far more common than a quest for excellence and accuracy complemented by tenacious follow-up. The answer is a management that demands truth and will do whatever is required to get it. Focus must be on process. Truth is the knowledge of things as they were and as they are. Not many accident investigation processes are designed to find it.

Accident Rate Goals

No one knows what it takes to achieve specific accident rate reductions. When there are goals to achieve specific accident rates, it is a dead giveaway that the management doesn't really know or, for that matter, care what is really going on and how the process is working. They just want to get the right answer or they really don't care what the answer is. Again, fear, pride, and selfishness (all negative emotions) take over and sincere caring for individuals is rarely even an ingredient. The solution here is to start with stopping accident rate goals. Only good things will follow.

Discipline Programs for Non-Compliance

Discipline programs take the emphasis off the complexities of teaching beliefs, values, and techniques to create a process free of defects. They encourage the easy out. Disciplining employees is like beating children for being disobedient. It may provide some satisfaction and achieve a short term goal but, over the long haul, does far more harm than good. That is not to say that the process should not include certain natural consequences for behavior that is inconsistent with organizational values. But people can't experience the natural consequences of behavior that is inconsistent with organization values until they have decided what their values are. The OSHA process encourages and models punishment. And OSHA offers little positive reinforcement. That is completely at variance with what is known about encouraging correct human behavior.

Workers' Compensation Cost Control

When costs are seen as getting too high, the all-too-common approach is to tighten the screws or get a new insurance company or broker. The best organizations know what their corporate costs are and have a responsible, early intervention method of handling claims that includes enlightened rehabilitation but doesn't focus on cost control per se. They know that if they do the right thing, they will get the lowest cost. The answer here is enlightenment. The most enlightened organizations are the self-insured, many of whom are members of the Workers' Compensation Research Institute in Cambridge, Massachusetts. If you join it, you'll take the first big step in fog abatement related to insurance/workers' compensation cost control.

One Stop Shopping Safety Solutions

Almost everyone is searching for the program, activity or training exercise that will "fix" their loss prevention efforts. Everybody wants the simple, inexpensive, quick answer, based on the appeal of instant gratification. The search is vain, doesn't work, sends the wrong message and inhibits real progress. Complex problems invariably require complex solutions. Accident prevention is a complex activity by its very nature. If managers don't do all they can to distill and simplify their approaches, accident prevention, like many other things in their lives, can overwhelm and confuse them. The answer is to eschew all "quick-fix programs" and rivet on process and a firm foundation, such as VIRO and values.

Incentive Programs

If individual rewards are given, one person wins and many lose; there are more unhappy than happy people. If the awards are by team, there are more unhappy than happy teams, and questionable team members may have inappropriate behavior rewarded and reinforced. If the reward is for everyone, anyone who has escaped injury by probability instead of optimal behavior has their mistakes ratified. These programs all focus on, reward, and encourage an emotional rather than logical approach to accident prevention. They are to loss prevention what snake oil is to healing. My abhorrence of incentive programs should not be confused with recognition for a job well done as it relates to process improvement, value refinement, and encouragement. "Atta-boys," however administered, are essential.

MINDSETS THAT ENCOURAGE APPLICATION OF LOGIC, SPAWN TRUTH, AND DEMONSTRATE CLARITY OF PURPOSE

The following describes several mindsets that are considered fruitful.

Fruitful Mindsets

- Root Cause Orientation
- Sincere Management that "Walk the Walk"
- Appreciation for Safety Excellence

Root Cause Orientation

An approach that diligently searches after and responds to root causes, with no expense or time constraints, will achieve excellent long term results.

Sincere Management that "Walks the Walk"

If every activity of management projects sincerity and reverence for their human assets, the organization is conserving its resources. Management should not only be willing to take the blame for losses but should seek ways to improve.

Appreciation for Safety Excellence

If a person doesn't appreciate or understand something (or worse yet both), the likelihood of their seeing it clearly is negligible. A management that has loss prevention values will understand how their organization is enriched by a sound and integrated safety process. Every conversation and act of that management will demonstrate their appreciation.

FRUITS OF DIFFERENT APPROACHES

The first thing learned in a basic logic course is that if A implies B and B implies C, then A implies C. The VIRO approach to loss prevention will encourage the same kind of analytical system that spawned the first laws of logic. Reliance on many other techniques can be likened to magic and smoke and mirrors. What you see is rarely what you get.

APPLYING THIS VALUE AND CLUES IT'S CATCHING ON

Many years ago, Jack Gausch developed a course based on clues to accident causation that was modified and later marketed by the ASSE. It was useful and inspired skills that could detect factors that had a high probability of leading to an untoward event. I'd like to provide you with clues to determine if this value—be guided by logic, not emotion—is being applied in your organization.

Clues of Being Guided by Logic, Not Emotion

The following page demonstrates several clues that indicate an organization is guided by logic, not emotion.

Clues that Logic Is the Basis of Decision Making

- Knee jerk reactions to accidents or losses are extremely rare.
- The results are good, but safety is rarely discussed as an isolated subject.
- Signs are few in number but, when required, are well maintained and explicit.
- Every manager can recite their primary exposures and how they cope with them.
- No incentive programs are connected to loss prevention performance.
- All employees know how the organization feels about safety and they like it and smile when they talk about it.
- Pressure for loss prevention is constant and consistent; there is no perceptible variance.
- Management is confident in their process. They are satisfied, but not self-satisfied.

APPLICATION TECHNIQUES

Applying this value requires vigilant introspection. Every person in a responsible position must frequently ask the questions: "Am I acting or reacting?" "Are we controlling the process or is it controlling us?" When the answers indicate that management is always acting and fine-tuning the process and in control, unhealthy emotion has been purged from their system.

14

EMPOWER OTHERS RATHER THAN SEEKING AFTER SUPPORT

"We give and grow."

· · · ·

"Not what we give, but what we share makes us great."

· · · ·

"It is not what we give, but how we give it that counts."

Perhaps the best article on safety management that has appeared in *Professional Safety* magazine in many years was "Safety Management: A Call For (R)evolution," written by my good friend and soulmate, Larry Hansen. The paper won the 1992 Irwin/Procter and Gamble Company "Best Paper Award" and then received similar recognition by the ASSE. This article is important because, like VIRO, it used the maturity concept but applied it to the entire organization and talked about only three states of maturity. The article emphasized that in companies who have "World Class" safety efforts or have reached his "Stage III" (the most enlightened) management had empowered the people. He pointed out that for too long safety professionals have hidden behind a "Dangerfield Complex," based on the famous actor, Rodney Dangerfield, who frequently had cause to say, "I don't get no respect." Larry probed the heart of what this value, empowering others, is all about. This chapter should add to the things he said, illuminate the subject to some degree, and set it within the context of VIRO.

RATIONALE FOR EMPOWERMENT

The discussion begins by exploring the nature of influence. Many of the people who have had the greatest influence on world thinking have had few or no titles and very limited resources. What they had were good ideas that enraptured people and captured their minds. Ghandi said, "My life is my message." He changed the lives of a quarter of the people on the planet, and influenced the lives of almost everybody else. Dr. Martin Luther King had a more profound effect on this nation than all but a few of the presidents. Einstein taught values as well as profound and insightful scientific facts that he didn't even bother to prove. Others proved them later for him. He was more than a brilliant scientist. Remember, he said that we couldn't solve problems with the same level of knowledge that created them. He obviously wasn't just talking about scientific problems. Einstein also said that people could choose to see everything or nothing in the world as miracles. He chose to see everything as miraculous. Christ, Buddha, and Mohammed all encouraged empowerment, had great influence, and did much good with few physical resources. By contrast, look at those who marshaled

power and depreciated human initiative, what they wrought—
Stalin, Hitler, Mussolini, Napoleon. What is the message here?
What model should society want to follow?

What do these great philosophers have to do with managing
safety and achieving good outcomes? Everything. They serve as
models for everyone and show them how to solve problems,
rather than ignore them, or worse yet...create them.

During a recently-completed project, a client wanted a
benchmark for what the companies who achieved "world-class"
in safety did to get there. The project was tedious until I started
encountering people who had already done it. I was already
ahead of some of those people, but one of them had spent far
more time and effort on the benchmarking than I was authorized
to spend and freely shared his results with me. Once again, it
was my good friend Charlie Dancer to the rescue.

During one discussion with Charlie, I tried to zero in on one
or two defining characteristics of those organizations who were
leaders in protecting their people, property, and the environment.
I asked if any one attribute seemed to stand out. The answer
iterated what almost all good safety professionals and line
managers know but may not focus on enough.

> *In **all** those companies, when you asked anyone,
> "Who is responsible for safety?", you **always** got
> the same answer: "I am."*

Making this response ubiquitous must be the goal for loss
prevention processes in search of excellence. There are many
routes, some requiring more or less time and effort, some
providing great side benefits like good scenery and growth

opportunity, and others bringing potential perils. VIRO is designed to provide a pleasant, efficient trip full of side benefits. This value deals directly with the subject and hopefully will add credibility to the VIRO process.

Ponder this analogy: The safety professional or any staff support person should be much like an official at an athletic contest. You will know they did an excellent job if, in reflecting on the contest, you can't remember them being there. If they ever draw attention to themselves, they have failed. You didn't pay to see officials; you paid to see a game.

Empowerment should be approached from all angles. The staff person must understand how to do it; management has to insist that they do it; workers need to expect it. If that is going to happen, everyone would do well to subscribe to Richard Eyre's new maxim in his book, *Don't Just Do Something, Sit There.* His maxim is, "Think stewardship, not ownership." This replaces the old cliché that "He who dies with the most toys wins." Stewardship should be defined as being responsible for the care and well-being of something that you don't own. In the case of the safety professional, it is implied that they are a care provider representing someone else (the management and the workers) and they will hold the safety professional accountable for how they discharged their responsibility. Some day the "safety consultant" will have to submit a "stewardship report."

Many readers would be inclined to disagree with his maxim, even after analyzing some of Eyre's logic. But this text will discuss the key points anyway, because they support this argument. Eyre quotes Henry David Thoreau when talking about the land he was surveying as being more *"his"* than it was the man who owned it because he *saw and appreciated* and took pleasure from it. Then he quotes Thoreau as comparing owning a farm to being in jail—the encumbering, enslaving aspects of ownership. I know some dairy farmers who understand this really well.

About stewardship, Eyre says, "When we begin to think we own our children, or our talents, or even our bodies, rather than perceiving them as gifts from God, we tend to value them less and to use them ever more selfishly." And, "Stewardship is a wonderful concept. It implies caring and responsibility without pride or envy." Finally, he suggests, "A steward doesn't necessarily want more. He wants to do his best with what he has. He wants more quality in his life rather than more quantity." Think about how ownership and power contrast to stewardship, influence, and persuasion. Everyone knows that power corrupts and absolute power corrupts absolutely. Sad experience has demonstrated that the nature and disposition of most men and women inclines them to exercise inappropriate dominion when they are given a little authority. The key to avoiding these pitfalls to progress is diffused responsibility or judicious empowerment.

EMPOWERMENT AT WORK

Perhaps this is a good time to finish the story of workers' compensation problems at Bass Shoe Company in Maine.

The president of the company, Rick Bourett, had told me in one of our early meetings that designing shoes was like eating a gourmet meal, marketing them was like having dessert, but manufacturing them was like the clean-up. He then likened his workers' compensation problem to taking out the garbage. I didn't agree with his analogy, but since he was using it, I suggested that he was drowning in garbage, and if he didn't do something fast, he would be in the garbage business rather than the shoe business.

Since Bass Shoe generated revenues in the low nine figures and had the potential to have a seven figure workers' compensation allocation of costs, I wasn't far from the truth. (For those of you who are counting and know what kind of profits shoe companies make, at that rate, losses eclipsed profits.) I rapidly convinced him that he should put his brightest and highest potential young manager in charge of the problem. Fred King got up to speed very quickly and developed the support systems he needed, and with the corporate resources we assembled for him to use, he turned things around in months.

Bass was no different than any other business. They wanted the fruits of their labors NOW. The corporation was suspicious. We had a complex web of brokers, fronting companies, third party administrators, service providers, and the premium allocations that went along with our captive insurance company. Under the existing program, they would pay for their misdeeds of the past for over four years. That's the way workers' compensation works. Rating is based on a three year period that begins twelve months ago. The process is very loss sensitive but has a long tail. Somebody can probably explain why it was designed that way, but I have never met the person.

There was a way out of this dilemma. Break with the current program and become self-insured. It sounds easy, but in this situation it was somewhere between difficult and impossible. A lot of paper is required, mostly in the form of money. You need a benefactor. In this case the only possibility was the parent company. Chesebrough-Ponds would have to stand up for Bass. Based on their history, our financial people were very suspicious of Bass' management and loath to risk company assets.

The last act began in the Board Room, once again. New characters included Bass president Rick Bourett, Fred King, the treasurer, and George Goebler, the COO. George was the senior person who would make the decision that Ralph and the Board would have to approve. Everyone got to state their case. Fred King projected that Bass could keep costs below $1 million next year. In addition, he made an impressive presentation of loss control and prevention activities designed to achieve his prediction. My claims manager, George Nelson, said that could be possible, but he would be more likely to estimate $3 million. The risk manager said that anything below $5 million was ridiculous and another runaway problem can't be excluded from the realm of possibility. He suggested that the risk was far too great for the corporation, based on prior performance. The division pleaded for the opportunity to take the risk and emphasized what they had done.

George Goebler asked what I thought. Another moment of truth for me. Do I stick with the corporate boys and make the saying "I'm from corporate and I'm here to help you," the greatest of the three great lies, or do I side with the division and have some limitations to the people I can eat lunch with in the cafeteria at corporate headquarters? I could eat out and travel a lot. I had twenty-four people who worked for me, so I figured I'd probably always have someone with whom to eat. Business relationships with the treasury department are important but doing the right thing is more important.

I decided to go with my instincts and my philosophy of empowerment, sink or swim. I didn't own that much stock anyway, and I have always been a gambler in terms of taking risks in my life. And, I knew Bass and Fred had a model process that was working very well. They had been good students and then good allies. I liked and respected them. So, I did it. I told George that if it was my money, I'd bet it on Bass. The decision came down days later that Bass was on their own and would receive the required corporate support. I had a few less people with whom I could comfortably eat lunch but a lot of friends in Maine. My business relationships would have to be the concern of my leaders who made the final decision. They knew what I had done and why.

At the end of the first year, Fred turned out to be wrong. His losses were under $300,000 and even fully developed (jargon for when you're all done paying five or ten years later) would be less than $500,000. When you set loose talented, highly motivated people and provide them with adequate resources, some very exciting things can happen. But, you must have the courage to do it and then to stand by them. I was proud that my leaders could do that.

There is one more Chesebrough-Ponds experience that illustrates this point.

Early in my career there, I developed a quarterly report that had bar graphs indicating safety performance. It was based on OSHA recordables and was an index that considered both frequency and severity. It was a variation on the old Z16 injury index that was the square root of the frequency times the severity. I don't think it was much better or worse than measures other people were using.

One day after lunch, I was sitting at my desk working and a figure appeared in my door frame holding up a copy of my report. I could see the bar graph and some writing in large letters but certainly couldn't read it. The figure questioned me, "Can you see this?" I wasn't even sure who it was, but I thought I recognized Gerry Chrusciel, who was the vice president for manufacturing for our packaged foods division. I had met him briefly before, but had never had a significant discussion with him. I had been to all his plants and knew his plant managers.

He told me who was writing on his report and what it said. He told me that the writing was that of his boss's boss, a Group vice president Bill Jackson, (at the time, the best candidate to be the next CEO). Bill was inquiring of Gerry as to why he always had the longest bar on the graphs comparing the divisions. Gerry then asked me why I was doing this to him. I was bold enough to gently remind him that I was only the scribe. I just collected and displayed the information. I didn't manufacture it. He said he had a few minutes before he was to meet with my division leader, Bob Bennett, and was wondering if I could tell him how to shorten his bar.

I said that I probably couldn't. He wondered out loud what use I was to him and suggested that it had probably been a mistake to hire me. I clarified my response and said that I thought that I could help him but not in a few minutes. He asked how long it would take and I said no less than an hour and no longer than two. He suggested that I call his secretary and make an appointment. I did. We talked for over an hour. He asked several good manufacturing type questions like, "What are we doing wrong?" I suggested that we didn't have enough time for me to answer that, but I could tell him what he was doing right in a few minutes. For some strange reason that seemed to get me off on the right foot with him. As with most manufacturing people (I like almost all of them), Gerry was a no nonsense person who liked to be talked to candidly.

Over the next several years, we talked a lot. He wanted to control his operations and I wanted him to be able to do it. He did what needed to be done and I helped him when he asked for help. We never became good personal friends, but we respected each other. He got to know what behaviors of his encouraged his plant managers, supported them, and correlated with improved safety performance. He learned them and practiced them with increasing effectiveness. Within a few years, he had the shortest bar. Then, for three straight years, one of his plants won the Chairman's Safety Award for being the best plant in the company from a safety standpoint. When I had been with the company for ten years, I got a nice service award...and a letter from Gerry. The letter was short and to the point. He congratulated me, expressed his appreciation for my help and the support of my subordinates, and said he hoped I'd be around for years. It doesn't get any better than that for a safety service provider.

What happened and why did it work? Gerry was a quick study. He wanted help, but he wanted it on his terms...and they were very reasonable terms; no nonsense and no frills. He expected no less from staff support people than from his manufacturing people. He only wanted cost effective help and quickly saw safety within the context of how it could assist him achieve his primary objective of turning out spaghetti sauces and meat tenderizers as efficiently as anyone in the world. Once he felt I could provide that, he **listened, believed, established his values, transmitted them to his plant managers, and insisted that they share his values**. He had been empowered and was empowering others. The truth is that after the first few years, we did less for his division than any other part of the company...but, he continued to be very successful.

HOW TO EMPOWER OTHERS

If an individual intends to empower others, he or she has to be prepared to give something up to get something. They will always get back more than they give, but must take that first leap of faith. One quality of love is that the more a person gives it, the more they have. Based on conventional wisdom, this seems illogical. Yet most people recognize this truth based on experience. Perhaps knowledge and influence are very similar but even less frequently recognized as being so.

Credibility is important. It's difficult to establish credibility and empower others, but there are some ways to do it. Before this section discusses that, I'll tell you how not to do it.

The Friday before I started my new job as corporate safety director, I was playing basketball at our swim club, the Cedar Hill Swim Club in Franklin Township. I was playing with some young men half my age, which was common practice for me. I was playing my normal "physical game" learned in the industrial leagues of eastern Pennsylvania. Someone got even with me instead of getting mad, and what initially appeared to be a sprained ankle turned out to be a broken fibula.

In spite of my protest of being a safety engineer and having a new job I had to start Monday, the doctor insisted on operating and screwing my fibula to my tibia. I called my new boss and told him. Ted Mullins wasn't thrilled with me, but was very considerate and told me to come in when I could. I was released on Monday and on Tuesday had someone drive me the eighty miles to Greenwich. I was in pain and on crutches but I needed to be there. Ted came into my office and told me that he would do anything he could to help me be successful. He then pointed to my cast, which was propped up on my desk to keep the pain level tolerable and closed by saying, "But you'll have to explain that yourself."

You know, thirteen years later, people still remembered the safety director who started work with a broken leg. I did largely overcome that—it wasn't easy, though. My debut certainly went against what I frequently told my children: "Avoid *even* the appearance of wrongdoing." I told everyone how it happened. Some people probably believed me. Falling off a ladder made a much funnier story, though.

The lesson to be learned here is if you're going to be a safety advocate, try to look the part. Above all, don't wear a cast to a new job.

This chapter will discuss some ideas as to how you establish credibility, but the person you *really* need to satisfy is your customer—operations management. Each group and individual is a little different. I will also later discuss what one enlightened safety services customer was seeing. That description will probably be even more useful to you than the ideas, but perhaps both together will be just what you need. Here are various ways you can establish credibility:

Habits to Form in Order to Be Seen as Credible

- Empathize and Understand
- Practice What You Preach
- Always Tell the Truth, No Matter How Much It May Hurt
- Be Subordinate
- Be Sincere
- Be Dependable
- Be Strong

Empathize and Understand

If someone thinks that you don't understand their situation, they are unlikely to accept any advice from you. If you've walked in their shoes, you have almost instant credibility.

Practice What You Preach

If you are ever found to be duplicitous; you are dead and so is everything you may be trying to do.

Always Tell the Truth, No Matter How Much It May Hurt

Once you are caught in a lie or even a distortion, you will never, ever be viewed in quite the same way. It is like a white cotton garment stained with indelible ink. With bleach you may get most of it out, but you'll never get rid of all of it.

Be Subordinate

Demonstrate that you are willing to subordinate your own interests to those of others. In brief, be selfless and not selfish.

Be Sincere

If you can't be sincere, then don't say or do anything. Judicious humor is essential, but people who joke around too much are not well-respected by most people, especially when the going gets rough.

Be Dependable

Never say you will do something that you can't or won't. If you can't come through, give loud and early warning.

Be Strong

Don't whine. Don't take advantage of those weaker than yourself. Help those in need. Be slow to compromise your beliefs and then only with good reason.

If you are not credible, you will have difficulty empowering others. You must have the *respect* of those you need to empower. As you empower them, they will enlarge you and the process feeds on itself. If you've experienced it, you know what I mean. If you haven't, try it; you'll like it. You can start by doing it with your children—think of them as your empowerment laboratory.

The following pages describe wrong ways for safety professionals to get respect or credibility.

Avoid Trying to Get Respect By...

- Doing *all* the training.
- Conducting accident investigations.
- Complaining.
- Using arcane language.
- Doing *anything* but earning it the old-fashioned way.

Doing *All* the Training

Some staff people seem to think that if they do a lot of training, they are essential and build their credibility by demonstrating all they know. This may work for a little while with a few people...but it gets tired fast. You become a noisy bell clanging in the wilderness. Your image is that of a tired preacher at the end of a too long sermon that people either don't want to hear or that fits into the "preaching to the choir" category. In either case, the result is bad. Some professionals will even create the impression that they are talking down to others. That sparks more than ambivalence. You become a human dart board. People start to take perverse pleasure in proving you wrong. Sometimes they will even set interesting little traps for you. The damage that can be done is limited only by their creativity and your sensitivity.

Conducting Accident Investigations

Your goal should be to arrive at a place where every operations person is expert in conducting accident investigations and transferring the skills to new people. All you do is consult on particularly difficult and complex situations and assist in interpreting results. That is empowerment. It isn't reached in many organizations. More commonly, the safety person either does the investigations or shepherds and nurtures them to the extent that they keep ownership. That's a bad idea, a sign that there is sickness somewhere.

Complaining

Many staff people—yes, including safety people—spend too much time complaining. They complain about inadequate resources, recalcitrant managers, impotent and disinterested leaders, misguided and wrongly motivated labor, uncooperative colleagues, and who knows what else. They think they're going to get sympathy, help, or be seen as a martyr. What they are most likely to get is lost respect. Nobody likes or respects a whiner.

Using Arcane Language

Every profession and discipline has its own language. Most professionals are insensitive to others who lack their specialized vocabulary. That's understandable. What is not, is when professionals *deliberately* try to posture and confuse people they should be trying to edify, help, and empower. It doesn't take long to decide with which you are dealing. There is a simple test. Ask the person using language you don't understand what they mean; then keep doing it until you understand. If the doctor, lawyer, accountant, etc., is posturing, they'll keep doing it and become irritated. If they are doing it by mistake, they'll apologize for confusing you and start to be more sensitive. People who try to gain respect by confusing other people with unique language will in time find their practice turning against them. They will lose credibility rather than gain it. The safety professional is no exception.

Doing *Anything* but Earning It, the Old-Fashioned Way

You earn respect slowly, through merit. There is no shortcut.

There are, however, numerous ways to legitimately earn respect. Here are some suggestions.

Earn Respect Legitimately By...

- Integrating your efforts.
- Empowering others.
- Enriching and challenging yourself.
- Following the suggestions of Mike Hostage.
  (*See below* for Mike Hostage's advice.)

Integrating Your Efforts

Look for ways to complement other objectives of your organization. Find out what goals other people have and help them achieve them. Start with people who report to the same person you do and branch out to those you serve and other colleagues. What comes around, goes around. You make deposits and you can make withdrawals. If you are a good person and good at what you do, there is probably no better way to demonstrate it than by working with people to help them attain their objectives. In the course of doing that, you will benefit in ways you couldn't imagine. Start small by learning how to do it. You'll like it, find it rewarding, and discover practice does make perfect.

Empowering Others

Search for ways to give people a piece of the action and then provide lots of "atta-boys." Express appreciation for what they are doing. Tell them how much it helps you. They'll feel good and enter into the positive cycle described above. You've just triggered it off from a different angle. Train the trainer is an obvious example. Show them how to do the safety training; incorporate their ideas; then let them go to it. Work with them on an audit program; participate in the first few; then let them do it on their own. Ditto for accident investigations and everything else you do.

PEANUTS   By Charles M. Schulz

PEANUTS reprinted by permission of United Feature Syndicate, Inc.

Enriching and Challenging Yourself

Do everything you can to facilitate personal growth. Express interest in anything that would help you serve the organization better. Be a hard worker and incorporate things you learn into what you do. Express appreciation for the opportunity to expand your horizons. Keep growing and keep thanking people for the chance to grow.

Following the Suggestions of Mike Hostage

So, who is Mike Hostage? He is the enlightened safety services customer referred to earlier. Back in 1976, Mike was the President of Marriott Restaurant Operations. He was the keynote speaker at the American Society of Safety Engineers 1976 Professional Development Conference. He spoke on what he expected of the safety professional that worked with him. His talk was turned into an article and published in the November issue of *Professional Safety*. I cut it out and put it in the top left drawer of my desk, vowing to reread it at least once a month until I had internalized the concepts.

Since that time, I have encouraged numerous people to read, study, and apply the principles taught in the article. It is a formula for success for any staff support person. But for a variety of reasons, it is particularly applicable to safety professionals. Mike started out by focusing on the importance of the relationship between the line manager and his staff support person. Then he discussed his expectations. They are exactly what any line manager would like but so often can't or won't specify.

The following lists the eleven attributes for which he was looking, with a commentary on each:

Mike Hostage's Desired Attributes of the Staff Safety Professional

1. Maintain Technical Competence in Your Field.
2. Be Aware of Your Own Technical Limitations.
3. Develop Good Communications Skills.
4. Be a Willing Resource.
5. Maintain Your Professional Relationships.
6. Be Skillful in Your Advisory Roll.
7. Be Loyal—Loyalty Is Always Admired.
8. Keep Key People Informed.
9. Make Completed Staff Work a Must.
10. Be the Eyes and Ears of Your Leaders.
11. Be a Beacon in the Darkness, a Social Lighthouse.

Maintain Technical Competence in Your Field

Participation in professional activities in *all* ways and at *all* levels is fundamental to excellence in all fields, including loss prevention. In this age of change, streamlining, expanded responsibilities, and reduced resources; broadening and extending the ability to contribute is essential. Then, the salient points must be artfully conveyed to leaders to avoid "fire-fighting."

Be Aware of Your Own Technical Limitations

Don't bluff on anything. It is easier for leaders to deal with subordinates that don't know and need time to research than with those who misrepresent, which destroys trust and credibility. The foundation of the relationship is damaged and will never be the same again. Maintain liaisons or lines of communication with top people in related fields so adequate resources are always available.

Develop Good Communications Skills

The loss prevention expert interacts with almost every sector of an organization (or should). That interaction is done in writing and verbally. Today, computer literacy is fundamental to effectiveness. Selling skills are essential—the ability to transmit ideas and persuade others to accept them will predict an individual's success rate.

Be a Willing Resource

If a subordinate is viewed as a pain in the neck, hard to handle, a credit grabber, or a threat in one way or another, that person will be isolated and ignored. If the subordinate appears to be pleased to help and enjoys the accomplishments of those they serve, line managers and others will line up to get their help. If clients/customers feel someone will make them look good and make their bosses look good, they will clamor for that person's support.

Maintain Your Professional Relationships

In spite of all the formal meetings and all the correspondence, casual personal relationships are invariably the basis of progress. Job descriptions and organizational charts seem to be needed in all organizations, but what really counts is how people relate to each other in spite of those artifacts, not because of them. If the informal relationships are understood and lubricated for accomplishment, you're an asset. Fundamentally, that will occur if those around you like to have you around—you make them feel good, secure. It is that simple.

Be Skillful in Your Advisory Roll

Here's where the arcane language and posturing can be deadly. Professionals must take the time to write the one page memo. They need to sense how much of something someone wants and needs to know. Meet the need. If their perception is blurred or inadequate, that's where the selling is important. Professionals must be willing to lose skirmishes to win wars. Sense when someone has heard enough. Watching the eyes is a good clue. When they start to wander or become glazed, you're done.

Be Loyal—Loyalty Is Always Admired

There are people who are in high places and stay there for a long time who don't seem to have any other attribute than loyalty. Be charitable of the faults of leaders and colleagues. Take the blame when subordinates struggle or fail. Pass on credit for successes. Be quick to compliment and slow to criticize. Compliment in public and criticize in private. Only write a critical memo or letter as a last resort...and then rip it up. Don't hesitate to place favorable impressions in writing, even laminated.

Keep Key People Informed

The caveat here is to have a heart. Recognize how much input with which people have to deal, especially those who are highly placed. Empathy and a keen sense of what is really important is critical here. Avoid the temptation to enhance people's impression of you by telling them *all* you know. Get to know audiences and adopt *their* priorities. Your priorities are of personal interest only unless you can influence others to share yours. Recognize that doesn't happen very often.

Make Completed Staff Work a Must

A reputation for never finishing things or accomplishing closure is stigmatizing. Keep a list and don't check off items until completed. Know the level of excellence expected by those you support and exceed it at all times. If you can't, get out. Remember, *what* a person does is important. *How* it is done is far more important. Neither is as critical, however, as *when* it is done. The old saying, "Timing is everything in life" may not always be correct, but a person won't go far wrong by adopting it.

Be the Eyes and Ears of Your Leaders

Leaders normally don't have anywhere near the exposure of their subordinates do. See the whole business and, without gossiping, share information that may be of concern to leaders. Staff support people should see themselves as extensions of the five senses of the people they serve: their customers, clients, and supervisors. When Ralph Ward asked me how a company president was doing, I was flattered. He listened carefully to my observations and asked good questions.

Be a Beacon in the Darkness, a Social Lighthouse

Everyone needs to be reminded of the boundaries at times. Long term considerations need to be weighed against expediency. Return on investment is always a consideration. The concerns of individuals can weaken or strengthen an organization, depending upon how they are handled. There are few people in a company who are better positioned to shed light on these issues than the person overseeing the conservation and optimizing of the physical and human resources of that organization. Being the chaplain can be very useful and enabling if one stays in touch with business realities.

These desirable attributes are as relevant today as they were almost twenty years ago when Mike Hostage counseled members of the ASSE on them. If a safety person can acquire them, I don't see how they can miss. The advice can, to a large degree, be applied by any support person.

Enabling and empowering cuts both ways. *Everyone* should be doing it to *everyone* else *all the time.* If it is not happening around you, catalyze it. Encourage others to do it. Demonstrate the rich benefits.

15

REFLECTIONS ON SUGGESTED VALUES AND OTHERS

"Reflection insures safety, but rashness is followed by regrets."

The ten suggested values and how they are presented should not be viewed as a completed work but as a work in progress. Each organization needs to establish its own set of values based on their circumstances. The characteristics of each organization will dictate not only what their values are or should be but how they go about determining and installing them. This chapter will ruminate on different cultures and industry groups and how hospitable they may be to VIRO.

The overall safety maturity of an organization will need to be considered when contemplating the applicability of VIRO. This text will lump companies into three levels of maturity; although further breakdown could be more accurate, it would make this chapter more tedious than necessary. But, this chapter will help readers enjoy customizing and personalizing. It has provided a maturity grid for safety processes. It was modeled after the quality management maturity grid in Crosby's *Quality is Free*. Managers could use it or a self-developed variant that meets individual needs. If you use it and want to see how your organization's safety maturity compares to its quality maturity, you can use the quality grid in *Quality is Free*. Such comparisons have been done with companies; they can be very enlightening.

There are obvious parallels/relationships between safety maturity and company culture. This chapter will review definitions and perceptions of the relationships between principles, values, virtues, culture, attitudes, and beliefs. Then, it will look at my suggested values and comment on how universal they are. Finally, it will suggest other values to encourage creativity, customization, and tailoring.

ORGANIZATION CHARACTERISTICS

Every industry has its own culture. And then there are subcultures. The nature and depth of cultures are very important when it comes to applying VIRO—how it should be introduced and how the indoctrination could be most effective.

Some years ago my wife and I watched a PBS documentary on Lee Iacocca. It not only provided insights into him as a man and leader, but also into his industry. When the program was over, my wife said, "I never appreciated the kind of language and crude behavior you had to be subjected to in your work." I was amused at her observation. She thought because meetings and interaction portrayed in the automotive industry occurred in a certain context, my company and indeed all industry behaved in the same way.

I pointed out to her that she need not feel sorry for me. In thirteen years at Chesebrough-Ponds, almost eight years at Merck, and almost four at Factory Mutual, I don't ever remember anyone using profanity at a meeting. I'm sure it occurred in private conversations, but then I just wrote it off as bad manners, intellectual atrophy, or in a few cases, creative thinking/more descriptive speech. I suggested that the automotive industry seemed to be more male dominated than where I had worked and appeared to have almost a military quality. I don't know if my observations have any validity, but my point is not to type cast any industry, only to suggest they are often very different in the ways they conduct business.

I believe the concept of managing safety or anything else by values would find the most fertile ground in creative and sensitive companies, perhaps those that place a lot of emphasis on internal human relations or external public relations through advertising and/or public affairs strategies. These organizations know how important their employees' input can be to what they are trying to accomplish. Companies like Disney, Marriott, Procter & Gamble, DuPont, Levi Strauss, and Weyerhaeuser immediately come to mind.

Organizations that are unionized routinely have problems with their labor relations and often even with their middle management and supervision. They have trouble understanding the concept of managing by values; they will find applying the concepts even more daunting. These companies or organizations won't be named, but everyone knows some of them. In order to even consider VIRO, they would need to do significant pre-work. But, this could be an opportunity for them to address and arrest endemic problems that have defied other solutions. They would have to start sooner, work harder, and spend more—unlikely without new leadership that was bent on positive change.

MATURITY

There are three useful levels of maturity that Larry Hansen suggests in his article from *Professional Safety*, "Safety Management: A Call for (R)evolution," that was referred to earlier. He calls Stage I "The SWAMP"—"Safety Without Any Management Process." These organizations are already experiencing excessive losses and resultant high insurance costs. They are afflicted with statutory ignorance and have adversarial employee relations. They will not understand much of this book. Teaching them VIRO would be like trying to teach geometry before covering fractions—not impossible, but not very easy or desirable.

Stage II—"The NORM, Naturally Occurring Reactive Management!"—is where most companies will fall. They will probably remain there...unless something happens to shock them into change. These companies don't really understand safety except in pockets; they perceive it as an uncontrollable cost rather than an opportunity to obtain a competitive edge. If they could catch VIRO, it would trigger off the "Radical Organizational Change" (ROC) that Larry says is needed for them to move to Stage III. These companies range from just out of the SWAMP to verging on Stage III, "World-Class." Obviously, the closer they are to Stage III, the easier application of these new concepts would be.

Stage III or "World-Class" companies see safety as a good business investment with long term benefits. In these organizations, line management owns and drives safety. The safety practitioners are few in number and high in quality, like the process and may have been the architects. One day when I was doing a project benchmarking "world-class" safety companies, I asked Larry to name the ones he had in mind. He could only come up with a few and admitted that his estimate of 23 percent in his paper may have been high. I agree; there are only a handful of companies who have arrived there. VIRO would only confirm for these companies what they already know. It would also provide an early warning if they started to slip. It would help them transfer some of their safety stardom to other pursuits. Applying a values-driven safety process wouldn't be very hard for these people; in fact it would be very natural for them. They already do most of what is encompassed in VIRO; they just have different nomenclature.

Using the Safety Management Maturity Grid can help define a company's level (see page 268). Different people/organizations will react differently to it, but it can only enhance introspection, almost always a good exercise.

DEFINITIONS AND RELATIONSHIPS

I have avoided defining terms earlier (a more conventional approach) because it would be undesirable to get hung up on definitions before there was a chance to unfold the entire story. If you've gotten this far, it is probably safe to talk about detail and get a little deeper into the theory.

Here are the dictionary definitions with an interpretation of how they apply to safety, VIRO, and each other:

Attitude: *A state of mind or feeling with regard to some matter; disposition: an attitude of open hostility.* The state of mind of a people is heavily influenced by their culture. How they behave is very closely related to their state of mind or how they feel. Behavior is comprised of acts; some of which produce good outcomes and some of which produce bad outcomes. People with "bad attitudes" seem to have more bad outcomes.

SAFETY MANAGEMENT MATURITY GRID

Measurement Categories	Uncertainty	Awakening	Enlightenment	Wisdom	Certainty
Management understanding and attitude	No comprehension of safety as a management tool. Tend to blame safety department for "safety problems."	Recognizing that safety management may be of value, but not willing to provide money or time to make it all happen.	While going through safety improvement programs, learn more about safety. Management becoming supportive and helpful.	Participating. Understand absolutes of safety management. Recognize their personal role in continuing emphasis.	Consider safety management as essential part of company system.
Safety organization status	Safety is hidden in manufacturing or engineering departments. Inspection probably not part of organization. Emphasis on appraisal and sorting.	A stronger safety leader is appointed, but main emphasis is still on appraisal and moving the product. Still part of manufacturing or other.	Safety department reports to top management; all appraisal is incorporated and manager has role in management of company.	Safety manager is an officer of company; effective status encourages reporting and preventive action. Involved with consumer affairs and other special assignments.	Safety manager on board of directors. Prevention is main concern. Safety is a thought leader.
Handling exposures	Exposures are fought as they occur; no resolution; inadequate definition; lots of yelling and accusations.	Teams are set up to attack major problems. Long-range solutions are not solicited.	Corrective action and communication established. Exposures are faced openly and resolved in an orderly way.	Exposures are identified early in their development. All functions are open to suggestion and improvement.	Except in the most unusual cases, exposures are prevented.
Cost of safety as % of profits	Reported: unknown Actual: 20%	Reported: unknown Actual: 18%	Reported: unknown Actual: 12%	Reported: 6.5% Actual: 8%	Reported: 2.5% or less Actual: 2.5% or less
Safety improvement activities	No organized activities. No understanding of improvement activities.	Trying obvious "motivational" short-range efforts.	Implementation of a program with thorough understanding and establishment of each step.	Continuing the program and starting to internalize safety.	Safety improvement is a normal and continued activity.
Summation of company safety posture	"We don't know why we have problems with safety."	"Is it absolutely necessary to always have problems with safety?"	"Through management commitment and safety improvement, we are identifying and resolving our problems."	"Loss prevention is a routine part of our operation."	"We know why we do not have problems with safety."
Safety culture	Almost no one knows what "safety" really means.	A few isolated people know that something is being missed when it	Many people recognize the virtues of loss prevention.	Most people support loss prevention efforts.	Everyone embraces the employment of preventive thinking and strategies.

Belief: *The mental act, condition, or habit of placing trust or confidence in a person or thing.* This is postulating that values are a direct result of what people believe. Once they believe in something, they place worth (i.e., value) on anything in their lives that support and are consistent with those beliefs.

Culture: *The totality of socially transmitted behavior patterns, arts, beliefs, institutions, and all other products of human work and thought characteristic of a community or population. Or, a style of social and artistic expression peculiar to a society or class.* Our usage focuses on behavior patterns, and the group is an extended family called the corporation. This book is postulating that beliefs and values will predict a culture. In this case, one that is not only hospitable to good outcomes but foretells them, feeds on itself, and oozes out into all areas of group activity.

Principle: *A basic truth, law, or assumption. Moral or ethical standards or judgments collectively. A fixed or predetermined policy or mode of action. A basic, or essential, quality or element determining intrinsic nature or characteristic behavior. A rule or law concerning the functioning of natural phenomena or mechanical processes. A basic source.* Stephen Covey has illuminated the dimensions of this word to a greater extent than any contemporary author. Covey says principles are like a compass. This is true. For a person on a ship for the purpose of taking a trip, a compass is essential. However, a sail or power source is needed to move the ship. Principles are the compass to direct individual values, and values are the driving force to that destination.

Value: *To regard highly; esteem. Or, worth in usefulness to the possessor; utility or merit: the value of an education.* If a people place a value on something, they will keep it and protect it. They will apply it to magnify it for themselves and those for whom they care. Their "values" are a grouping of those things that are important to them and which profoundly affect the choices they make. "For, where your treasure is, there will your heart be also" (Luke 12:34).

Virtue: *Moral excellence and righteousness; goodness. A particularly efficacious, good, or beneficial quality.* In putting together the best selling book, *The Book of Virtues,* William J. Bennett has focused attention on this word and the importance of virtues. In *The Moral Compass* and a book on virtues written for children, Bennett continues to illuminate the importance of our forebearers' emphasis on correct behavior through reprinting and commenting on the writings of numerous authors, including many classics and some obscure treasures. Virtuous behavior will come from holding values congruent with correct principles.

Society has experienced so much promiscuity that the word virtue is losing much of its traditional meaning as it becomes associated almost exclusively with female chastity. Its true meaning is far more comprehensive. Virtue has a place in values-driven safety. Applying VIRO *is* virtuous, by definition.

REVIEWING SUGGESTED VALUES AND THEIR UNIVERSALITY

Do It for the Right Reasons

This could be stated in different ways. Not valuing purpose is at the root of many failed safety processes, or at least associated with the inability to perpetuate them. The end cannot justify the means in many cases. Improperly motivated activities, in any area, will not have the success rate of properly motivated ones.

See It as Part of the Whole

One can be successful at accident prevention while viewing it through the eyes of a "Humpty Dumpty" manager. Many companies achieve "world-class" without seeing safety "as part of the whole." However, VIRO is a higher law, will differentiate a loss prevention process, and can make extending efficacious values-driven processing to other areas much more natural.

Recognize There Is No End

Spasmodic, lurching movements in any activity rarely produce a harmonious and consistent result. This is true in safety. If good outcomes are achieved without a long-term, sustained effort, it is probably an "accident"; there will be trouble replicating it—except randomly. However stated, this concept is central to safety process success.

First It Is a People Business; Things Are a Distant Second

Several very successful safety programs are driven by engineering. That is the primary thrust in the United Kingdom. By contrast, the whole concept of values-driven safety focuses on people. If this is not stated as a value, it needs to be well understood. This *could* be so obvious that it can go without saying...but, it is doubtful. As with the other values, its applicability extends well beyond safety (consider customer relations, for example).

Put the Right Person in Charge

No process or activity can rise above the quality of its leadership. With safety, the correlation is strong between the person (the safety professional and/or their supervisor) and the product (the losses). This value is essential for all organizations. It could be worded differently, but it would be difficult to abandon this concept or belief. Switch the conductor and the bass drummer—hand over baton for mallet—and you are unlikely to get very good music.

Use a Yardstick that Everyone Can Read

Measurement is central to any success. Drucker stated a truism: "If you can't measure, you can't manage." This could be said differently or incorporated in a broader value, but it is important enough to be set apart.

Sell Benefits...and There Are Many

Generally, if you don't sell something, it doesn't sell—in spite of its merits. In a very mature safety process, aggressive selling may not be necessary. *Perhaps* benefits need to be sold in the beginning, but at some level of maturity, dropped...but, it's doubtful. Selling benefits could simply become very subtle or may just take place as a natural result of other beliefs and values.

Never Settle for Second Best

I may have a little Rodney Dangerfield in me, too. But not much. This may take care of itself in the natural course of events. It may be better left unsaid, as it may create more wall building than bridge building. It could encourage defensiveness. Managers should think a lot about this one before installing a value-directed loss prevention program...but, I doubt I'd delete it.

Be Guided by Logic, Not Emotion

The concept behind this value can be phrased many ways. It would be okay to experiment with the language, but the message should be retained. Emotionalism and irrationality are obstacles in almost every problem solving exercise. Bad outcomes are almost always associated with decision making flawed by unfounded biases...perhaps a euphemism for negative emotion.

Empower Others Rather than Seeking after Support

Although this value is written to address safety professionals, it applies equally well to other professionals. (Empowerment aids safety and other business spheres.) Smoother and more descriptive language could undoubtedly be crafted for this value. But, you get the point. Many hands make light work. Everyone knows that safety doesn't work unless they get buy-in and almost universal participation. This value deserves distinct billing and maybe even top billing.

OTHER VALUE OPTIONS

Although I'm somewhat wed to my own list, the following are more values and variations intended to stimulate the reader.

Talk Safety On *and* Off the Job

This is a little narrow and specific, but it is important. Most of the companies who have achieved "world-class" believe it...and they do it.

Make Safety a Condition of Employment

This is direct and maybe even too harsh. VIRO takes care of this value inherently. Managers should state their beliefs with regard to safety. If a prospective applicant doesn't share these beliefs, they'll hopefully find employment elsewhere.

Ultimately, Management Is Responsible

But management can't take all the responsibility away from the employees. Unlike my ten values, this value should be felt by management but not promoted as a company-wide value. It could too easily lead to misunderstanding. It's the other side of "witch hunting" and "blame placing" that undermines so many safety efforts.

React to Every Incident, Not Just Accidents

The overarching VIRO concept should supersede this true, but somewhat impractical, value. We could do without this one. But...we're just brainstorming, right?

Preventing Injuries and Illnesses Is Good Business

Several of the suggested values include this concept, but for some companies, maybe it would be best to state it this directly.

Safety Must Be Designed In

This lends itself to a lot of interpretations. It could focus on proper design of new and renovated facilities. Or, the aspect of Job Safety Analysis review that uncovers and addresses hazards introduced with process or job modifications could be targeted. For the process industries, this value could just focus on process safety. I like it...even with its ambiguity. I think I could write a good chapter on this one.

All Injuries and Illnesses Are Preventable

This is theoretically true. Practically it is not. This statement (and we've seen it) inspires too many arguments. I never liked it. None of us should ever accept bad outcomes. We need to believe we are in control and what we harvest will be what we sow. But I can find a lot better ways to say this. On the other hand, if this thinking is part of your culture already, it shouldn't hurt to formalize it as a value.

CONCLUSION

Hopefully, this reflection will help readers think of these as just "suggested values" and prompt you to modify them and/or come up with your own. This book is not intended to sell specific values, but to suggest a way of thinking. The next step is to describe practical applications, which is where the pay-off *really* is.

SECTION III

IMPLEMENTATION

AND MEASUREMENT

16

ESTABLISHING THE CONCEPT IN AN ORGANIZATION

"Change of pasture makes fat calves."

. . . .

"Don't let fear hold you back."

. . . .

"Fear is a great inventor."

INTRODUCTION

In order to establish the VIRO concept in an organization, there must be a clear understanding of the theory and application possibilities. The management, the employees, and the safety professionals all must grasp and support the premises. They must also understand the concepts and be anxious to enlarge them and tailor them for the organization of which they are a part. The first two sections of this work provide the theory and *some* application *possibilities*. Finally, there must be a realistic grasp of the details associated with the preparation, tools, blueprint, and execution of the exercise. The next two chapters will provide those nuts and bolts.

If the process can be installed in only a part of an organization, it can, at best, only come close to achieving its full potential. Clearly, the almost infinite organization configurations and relationships affect the efficacy of VIRO. If an organization is made up of almost totally autonomous groups, doing VIRO in one of them may work *pretty* well. In a centrally controlled organization, trying to do values-directed safety in a part of the group would have little or no impact.

LEADERSHIP ORIENTATION

The best case scenario would be for the CEO or COO to be enraptured with VIRO (the concept). Someone like Bob Haas of Levi Strauss, who is already managing his company by values, could get very excited by this method of managing safety. Chapter 19, Extended Applications, will talk more about what he is doing at Levi Strauss. A credible and influential safety professional, one who is enthusiastic about the potential of this concept and wants to make it happen, will hopefully marshal top executives and develop a strategy to accomplish it. The details—simple to install in a setting of enthusiasm and support—are discussed throughout the book, with the heart in the next two chapters.

Selling VIRO depends on the environment and what has worked in the past. Everything the leaders and safety professionals do must be culture and organization compatible. The simple way would be to convince someone who has the ear of the leader and get them to be the champion. If several of those kinds of people in different parts of the organization chart can be convinced, so much the better. Under certain circumstances, this could happen in weeks. In most, it will take months or years.

At Chesebrough-Ponds, the forum might have been the Corporate Safety Committee made up of the heads of manufacturing for all the companies. I might also have had some success by getting a Group Vice President interested, and in turn, company presidents. I would also have laid the ground work with normally supportive staff groups, such as public affairs and human relations. They would not have driven the process, but would have spoken up when asked. I would not have tried to start with the company presidents because they were marketing-oriented and normally had little to do with safety. They responded to what their leaders and manufacturing heads suggested to them.

ROLE OF THE SAFETY PROFESSIONALS

If the leaders of the environmental safety and health movement are caught up with the concept but no one else in the organization knows much, if anything, about it, there is a whole different story. If those people lack influence, VIRO can't go anywhere. If they have sold new approaches to loss prevention in the past, this could be their biggest challenge yet. It could make them heroes in their company and ensure their job security for many years.

No matter how this is sold to a company, it can't materialize in its full glory without a strong, effective, broad-based, talented safety practitioner in place or injected early in the effort. It also can't happen without an independent outside facilitator who is intimate with the process and the field of loss prevention by its broadest definition.

Both the line management leadership and the staff support persons inside and outside the organization must exude an almost religious zeal. They must be willing to work selflessly to bring others on board and teach them the concepts; evolving the values, measurement, and edification as they go; giving credit freely to anyone who deserves it.

"*Now that desk looks better. Everything's squared away, yessir, squaaaaaared away.*"

EMPLOYEE INVOLVEMENT

The workers will be the least of the problem. They already know what is wrong with the process. They will love this just like the people at the Health-tex factory in Portland, Maine took instantly to my chiropractor friend, Pat Venditti.

The sooner the workers get involved, the quicker the process will advance. They will intuitively grasp the theory and relate it to their families and clubs. If given the opportunity, they will drive it like nothing else in which they have ever been involved relating to safety. They will see this as a "put up or shut up" opportunity for them to get management to do what they say.

GETTING IT ALL TOGETHER AND GETTING ON WITH IT

The whole process will scare insecure members of management. The weak ones have every right to be afraid. They are about to be exposed...wherever they are. *Fear* could be an impediment or a driver, depending on how it is handled. *Change* could be viewed as threatening or enabling, providing new opportunities.

This is a short chapter because it is hard to envision every corporate climate or organizational structure, and specific approaches to establishing the concept will therefore vary. But this is one of the fun parts, since it is so challenging and each situation so unique. If you can't establish the values-driven concepts, it doesn't matter how useful they can be. If the idea is as transforming as it seems; why not go for it? What have you got to lose? Be bold!

17

INSTALLING THE PROCESS

"Little by little the bird builds his nest."

Installing VIRO will be a journey. Management must have a vision of where they are going and of the full potential of the exercise. If they know where they want to go and are motivated to make the trip, only one other piece of information is required. They need to know where they are now in relation to this process. Hence, the first step is an orienting exercise. Most people and companies don't really know where they are. The maturity grids exposed in the next chapter should be used as part of an initial assessment.

WHERE ARE WE?

The Program

The organization should start with a comprehensive, more or less traditional evaluation or assessment of their safety and health efforts. If one has been done recently, it may only need to be updated and focused on what is relevant to the installation of a values-directed approach to loss prevention. The orienting will, in effect, be a comprehensive measurement of all aspects of the efforts to minimize exposure and loss. All efforts should focus on areas that are compatible with the concept of managing safety by a set of values and shun activities that could undermine or stall the philosophy or the mechanics of its installation.

Major helpmates would be self-directed work groups, behavior-based safety teams, harmonious labor relations, recent culture change activities that were successfully executed, robust business, success in implementing a TQM thrust, significant executive interest in principle-centered leadership, disenchantment with traditional safety "stuff" that is more of an impediment than an enabler, and strong internal training resources.

Trends counter to the above would need to be reversed if a values-driven approach to loss prevention is to be successfully achieved. In addition, the following, if present, must be overcome for successful installation:

- Strong bias toward either decentralization or centralized management.
- An arrogant and/or inbred corporate culture.
- An insecure safety group who are defensive about new ideas.
- A safety group that does not enjoy universal acceptance by operations.
- A resource-starved safety effort.
- An environment that is overly bureaucratic.
- Discredited or weak leadership.
- Excess influence by finance or legal departments.

This discussion will start at the bottom and work its way up. If current efforts are regulatory and/or insurance-driven due to very influential executives, this effort could fall flat as turf wars erupt all over the place. There are plenty of lawyer jokes and many about accountants. Two personal favorites will help to bring bittersweet humor to this warning.

A lawyer, doctor, and engineer are found arguing about which of their professions had the earliest origins. All three of them are in agreement on one significant point that often divides people. They all subscribe to the rendition of the creation in Genesis as opposed to the theory of evolution. The doctor starts his argument by referencing the creation of man and woman by God in his own image; clearly a medical miracle of the earliest origin. The engineer couldn't wait and waxed eloquently about the creation of the earth prior to that activity; clearly an engineering job. He pointed out that at first, there was darkness and disorganized matter—in chaos. Organizing that chaotic matter had to be done by an engineer. The lawyer at first was quizzical; then smirked; then began to grin...he had it. Simultaneously the doctor and engineer asked: "How can you top that?" His answer came in the form of a question: "Where do you think the chaos came from?"

Let's not blame the lawyers, though—it's their training. In school, they are trained primarily to find problems. Not much time is spent in solving them. Most never overcome this education fault.

Even some comments from Biblical times on lawyers are generally relevant today. Jesus Christ said, "Woe unto you also, ye lawyers! for ye lade man with burdens grievous to be borne, and ye yourselves touch not the burdens with one of your fingers" (Luke 11:46). And, "Woe unto you, lawyers! for ye have taken away the key of knowledge: ye entered not in yourselves, and them that were entering in ye hindered" (Luke 11:52). Lawyers apparently were educated the same way 2,000 years ago as they are now.

Installing VIRO requires everyone to roll up their sleeves and pull together. It is, in a sense, a spiritual activity in that people need to delve into themselves, find out what they really believe, and determine how that translates into what they value. Lawyers can't help much, rather they are naysayers. There is a saying about naysayers.

Eckenfelder's Rule of Animate Objects states, "Naysayers litter the landscape; they are like dandelions before the discovery of broad leaf weed killers. It is best to just relax, ignore the dandelions and enjoy their bright yellow color."

Let's contemplate accountants and how they may be able to *help*.

One day, a Frenchman who was an avid hot air balloonist jumped into his gondola, fired up, and took off. A rapidly developing wind out of the east blew him across the English Channel, then across England itself, and finally across St. George's Channel. He landed in a field in Ireland. In Ireland, cars are relatively expensive and most people ride bikes, even professionals. A man with a smart tweed suit and an expensive leather attaché case was riding by as our balloonist landed. He dismounted and walked out into the field. He was greeted by a dazed man draped with a balloon. Haltingly, the man inquired as to his whereabouts. Our biker answered, "You are in a field in the middle of Ireland sitting in a gondola, with a balloon draped over your head." The Frenchman became alert and in crisp English with a French accent, he observed, "You must be an accountant." The cyclist was at first amazed at the accuracy of the observation but then quickly appreciated *his dress, case and demeanor that might have given him away. But he instantly reasoned, why not a doctor, engineer, or just a successful businessman? His rejoinder was swift: "How did you know that?" He played the perfect straight man. The answer was instantaneous: "What you have told me is well-intentioned, precise and accurate in every detail, yet completely useless to me."*

In most organizations, if there is a claims function, it reports to the finance group—a mistake in almost all cases. When I hired George Nelson to be our claims manager at Chesebrough-Ponds, he was largely ignored by the risk manager, treasurer, and their colleagues and bosses. When he started getting a lot of recognition, they wanted him. I thought it would be a good idea to have a friend in their camp and decided to let it happen. But, first I needed to know what they would do with him. I asked them to tell me. They did and I changed my mind. They intended to have him spend most of his time in the office compiling information for reports and renewals.

I said no, and made it stick. With us, the loss prevention department, he was mostly in the field helping us avoid claims, and when we had them, got involved early to mitigate loss. (My *loss prevention after the fact.*) I wanted him to help us win. They wanted him to keep score. When it comes to VIRO, activity is play-to-win, not just keep score. Bean counters are little needed.

Developing widely accepted and supported values and concepts will require that the senior executives are universally respected; otherwise they won't be able to exercise the moral leadership required. In addition, the greatest asset that an environmental safety and health group can have is the trust of the people they serve. If operations management sees the safety people more as a part of their problem than a part of their solution, the safety professionals are headed to uselessness. Such a posture cannot lead a renewal and energize safety. VIRO *is* a whole new idea. If the safety department is insecure and needs to author all safety changes or goes around feeling badly because they weren't invited to meetings, they will have trouble with values-driven safety. If *they* have trouble with VIRO, trouble will follow for everyone because safety staff support will be needed more than ever when values-driven safety catches on and the scope of their involvement is expanded. If they can't see the process catching on, they may have incurable organizational myopia—unlike regular myopia, this is ultimately fatal.

In today's world, we can't stand still and must be adaptable. An ability to view loss prevention in a new light is essential to applying VIRO. The only thing good that video games teach children is that if they don't move, they are dead.

If the potential host environment is driven by procedures and paper, the culture will choke on values-driven safety. Properly integrated, it does not need much paper. In fact, a group dependent on paperwork as a security blanket may become disoriented by VIRO's simplicity and withdraw.

Recall the dictum, "You have to spend money to make money." Similarly, you have to spend money to save money by avoiding unnecessary loss. The beauty of the loss prevention process, though, is that the return on investment is invariably much greater...and when it is done, there is a leaner, meaner, more competitive enterprise. If an organization has been in the habit of starving their safety efforts, they may struggle to understand how these concepts actually work. It's a little like trying to convince someone who keeps their money in a mattress that the bank would be a better place for it.

A keystone is that wedge-shaped piece at the crown of an arch that locks all the other pieces in place. After securing the bases, putting pressure on the keystone can actually strengthen the arch. The keystone of VIRO is the concept that beliefs and values will predict outcomes and that they can be taught. A proverb states, "A man's *actions* are motion pictures of his *beliefs.*" Installing the program keystone will be tricky, but once properly in place will only strengthen the arch if pressure is applied.

While VIRO is not driven by traditional religious values, people who hold to some set of religious beliefs and the accompanying values will be more comfortable with this concept because they will be able to draw analogies and understand better. Virtually all religions encourage humility, perhaps because it is the path to learning. If you know everything or think you do, what can you be taught? Trying to "plant" values-driven safety in an organization with arrogant, inbred, and self-satisfied management would be like sowing seeds on concrete.

If a company is too decentralized, it is not accustomed to corporate leadership. Using the family analogy, if the father doesn't go to church, the children probably won't either. If there is no father or if the father is never home, there will probably be the same result. If management is centralized and autocratic, the subunits probably won't be strong enough to fulfill this goal. A father who does everything for his children is unlikely to produce resourceful adults. Religion can't be imposed on children. People can't be told what to believe. They must do it to themselves; no one can do it *to* them.

The bad news is that these organizational obstacles provide challenges that may derail VIRO. The good news is that if they are overcome, not only will VIRO benefit the organization, but it will be much healthier for the experience.

SUPPORT SYSTEMS

If a crop is planted in an arid climate and has no irrigation, the farmer has wasted his time and money. Likewise if he has not planned harvesting. That is even worse because more effort has been wasted. What support systems are needed to feed, water, and harvest VIRO?

1. Safety professionals need to understand the value of training and have the internal resources or be prepared to access needed assistance.
2. An existing solid traditional safety program that wants to move to a new plateau is helpful and suggests a capable safety staff and training resources.
3. The people who know a lot about employee assistance and wellness programs will relate immediately to VIRO concepts. They can really help.
4. If the senior executives are amenable to venturing outside their *comfort zones*, have displayed a willingness to support off-the-job safety, and have developed the habit of speaking out on social and community issues, they can drive VIRO like nothing else.
5. The human resources people can be an enormous help if they are capable and broad-minded enough to see what VIRO can mean to them. If they can't and get defensive, storm clouds may encroach and possibly rain out the parade—if they are influential.
6. An outside facilitator will be essential. Positive past experiences with consultants would be very helpful. If management and/or employees are antagonistic toward outsiders, it is unclear how VIRO can be installed. A very competent staff person could accomplish the mechanics of

this process, but remember, "No prophet is accepted in his own country." So purely inside attempts at installation could short fuse.

Developmental History

Past experience in a company with self-directed work groups (or whatever you may call them) can be preparatory. I saw the beginning of work groups without traditional supervision at Chesebrough-Ponds as an appendage to Just In Time efforts we started almost ten years ago. I don't think I had much of a grasp of what they were all about except that the concept felt good. I was a little puzzled as to why they hadn't caught on more in this country and, more to the point, why they seemed to fail so often.

The lights came on for me at a brief presentation made at my local Adirondack Chamber of Commerce management development meeting. Chandler Atkins, who teaches at the local community college and does some consulting, synopsized the process and the problems. Chandler is an insightful, Renaissance man who dabbles in many fields, including politics. He has the scoop on self-directed work groups, and his explanations have led me to see parallels between installing employee involvement manufacturing and VIRO. The points of commonalty are significant and relevant enough to warrant consideration here as we discuss installing the process. If someone has successfully installed self-directed work groups, VIRO installation will be a lot easier. If they haven't and want to do so, VIRO could help pave the way. The common points are found on the following page.

Common Points Between VIRO and Self-Directed Work Groups

1. The processes must be seen as more important than the goal.
2. Preparation is critical. Feasibility must be carefully evaluated.
3. The proper environment is essential. There are go and no-go clues.
4. Universal concurrent involvement is called for.
5. Training will always be a major ingredient.
6. They are first and foremost people processes.
7. You build on what is there. (You get to cut through irrelevant documentation and bureaucratic "red tape" and see what is *really* important.)
8. Communication is essential.
9. There is no end... And that is good.
10. They have advantages and pitfalls.
11. Maturity is at the center of both.
12. Benefits must be sold hard until inertia is built.
13. Leaders must be *totally* committed.
14. Core benefits are striking; side benefits unimaginable; and total benefits uncountable.

1. Focus needs to be on the process rather than the goal. In a goal oriented society, this concept is often hard to understand and even harder to apply. Goal setting may not always be all it is cracked up to be. People are often counseled by goal setting gurus. Your goals must be tangible and measurable. Having a goal to be a better person is a good idea, even if an individual can't hold it in his or her hand. This brings to mind the very popular Job Safety Analysis program, which is experiencing a revival. Most people who do it see the finished analysis as the product. However, the tangible product is clearly a by-product. The real products are the dialogue with the employee during the preparation and the discovery of hazards, the training for all participants, and the impetus to review job changes in order to discover new exposures introduced.

With self-directed work groups, the product is an enriched and enhanced work force, not more and better product. With VIRO, the product is not fewer accidents; it is an enriched and enhanced work force with a by-product of fewer losses.

2. Without adequate preparation and training, self-directed work groups will fail. If not prepared, employees cannot manage themselves, much less someone else. Getting people ready for culture changes like these is complex and time-consuming, yet essential. Solid preparation is essential to successfully install VIRO.

3. The ground must be fertile and prepared for "planting" of both processes. A list of go and no-go clues can and should be developed. For self-directed work groups, Chandler provided these examples:

GO CLUES:

- The top executive champions employee involvement, encourages dignity and trust throughout the company, and acts as a mentor to other potential leaders.

- The company is willing to stick out a two-to-five-year transition to mature teams.

NO-GO CLUES:

- Mid-managers uniformly and actively oppose employee involvement.

- Insufficient capital is allocated for essential training and technology.

VIRO has a different set of GO and NO-GO questions, but they should be developed and used to avoid predictable failure. However, *unlike* self-directed work groups, the cost will be low and even partial results will have some value. But, you should start out unsatisfied with a part loaf. Why not have it all?

4. Unlike some programs that can bubble up or trickle down, these two need to start on all fronts at approximately the same time. They require a full effort to overcome negative inertia, and the desired reaction can be stopped, or at least inhibited, by only one sour ingredient—right?

5. The list of the information dissemination and skill building required for either of these efforts is substantial. A significant and capable training resource is essential.

6. In both cases, people will determine whether the company succeeds or fails. Managers can't *buy* either of these processes. Selecting and training people who will not only accommodate the ideas, but who can become drivers themselves is delicate yet essential.

7. Companies cannot start from scratch. They must start where they are and build from there. In the process, they get to see what is really important to their success: their people and their ability to get the best from them.

8. All types of communication at all levels are central to VIRO or self-directed work group success. And not just the communications experts need to be good. Everyone must communicate, and this will be talked about more later in this chapter. Building communication skills will almost surely be necessary.

9. As suggested in the third value (see Chapter 7), VIRO itself has no end and neither do self-directed work groups. They become a way of life and are always being enhanced. That is their real beauty. You keep planting but you keep harvesting.

10. These are organization changing efforts. They have wonderful benefits but also numerous pitfalls. Unlike self-directed work groups, which can cause business disruption or even failure if not implemented correctly, VIRO has no dire downsides.

11. Both of these organization/process change efforts go through a maturation process. That process can be traced and measured. Measurement is essential, and this has been explained in Chapter 10.

12. Value Seven—Sell Benefits, and There Are Many (covered in Chapter 11) discusses the benefits of VIRO and loss prevention and how they need to be sold constantly to achieve optimal performance or very few bad outcomes. Analyzing why work teams fail almost always uncovers deficient selling of the concept.

13. Leaders can't blink with either of these programs. If they blink, they lose. It's like the game children play, staring at each other until the loser blinks.

14. If either program is done well, the benefits almost defy complete description. The organization opens up a whole new world for itself. They probe the depths of their greatest asset, their human resources, and in so doing unearth hidden treasures.

HOW ABOUT OUR CULTURE MATURITY?

Culture/History/Background

The initial surveys will expose the history of the safety movement in the organization and provide the background needed to move ahead. The safety culture will become very evident. Hence, VIRO *will be culture sensitive right from the beginning* and the values-directed process will just grow and deepen. Other approaches to safety are culture blind, as are most approaches to making this a better world. That is why many of them fail. Look at crime, equal opportunity employment, welfare, education, and health care. Most of the improvement programs are culture insensitive.

The VIRO Maturity Grid

To start the process, cultural maturity as it relates to safety must be assessed. The discussions earlier in this chapter covering latent culture and parallels with self-directed work groups could be viewed as traditional. Now safety professionals need to embrace the belief/value driver concept and decide what values will drive their organization to safety excellence. Once they have done that, they need to define levels of maturity and exactly how they will see how mature they are. One definition of personal maturity says that someone is mature if they are willing to do something they don't want to do at a time they don't want to do it and still do it as well as they can. The maturity tables suggested in the next chapter imply total maturity when the organization has done all it can, all the time, even under stressful conditions.

The maturity grid is the heart and soul of VIRO. This grid could be used effectively in most organizations to vastly improve their outcomes. But, hopefully organizations will want to develop their own. The development process will facilitate every aspect of VIRO.

GETTING RESULTS AND INTERPRETING THEM

Using the Maturity Grid

Once an organization has decided that VIRO is something they want to do, maturity grids take center stage. They will serve as the baseline or reference standard. If a good job is done in preparing them, and if they are on target for the organization and weighted to be most predictive of the outcomes desired, they may never need to be changed or may require only slight modification.

VIRO offers substantial flexibility as to just how precise to be. Certainly, almost any efforts made will enhance the current approach. If safety professionals do a lot and pour themselves into it, who knows just how great the results can be? If they just want to use the prototype in the book, they will be pleased with the result.

Either extensive or no evaluator training can be done. A positive result will occur either way. The dividends just increase exponentially if greater investments are made. The investments are primarily spiritual, but also financial, as a side light. The primary tools, the maturity grid and user instructions, are modeled in the next chapter, with their use and options explained in more detail.

Interpreting the Results

Organizations can go as deep as they want with interpreting results of the maturity grid surveys. If they just poll twenty people (that is the minimum to get a representative sample) and only do it once, they will learn a little. If it is a very large and complex organization, one that does hundreds of grids and has lots of statistics already on safety, the survey will be very revealing and could yield conclusions for a very long time. The interpretation exercise can either be simple and quick or complex and brain straining. Each company has the choice and can grow or shrink their efforts as they please. Doesn't this all sound more engaging than deciding from whom to buy safety shoes or how to comply with the latest OSHA regulation? *More importantly, it will produce far greater long term benefits.*

THE NEXT STEP—AFTER ANALYZING AND INTERPRETING INFORMATION FROM THE MATURITY GRID

The Vision and Its Importance

If you are not excited about this concept, don't get started. I don't want VIRO to fail. It can't fail if you're excited about it.

The way to get excited is to develop a vision of where it can lead you. I get excited about new ideas that can effect positive change when I ponder them; then I dream about them; then I daydream about them; then I tell everybody about them—that binds me to a contract I've made. Then I have to do something about my talk—whether it means buying a new boat or building a house or just losing a few pounds. The secret is to first get a vision of where you want to be and how good you will feel when you get there.

Over forty years ago, a wise baseball coach taught me about visioning. It propelled me from being an average Little League pitcher to what a lot of people said was the best high school pitcher in New Jersey in 1958 and a Group IV State Baseball Championship. All my children have worn the championship jacket, and it still hangs in my son's closet, albeit a little faded and tattered. It was Teaneck High School. That coach was Art Degerick, and he was way ahead of his time. I owe any recognition I received to "Mr. D." and the things he taught me. Visioning was perhaps the most important lesson.

Only in the last decade or two have sports psychologists taught the value of visioning for winning performances and mentally erasing failures. The same thinking can be extended from sports to all other activities. I know it works because I've done it. I'll bet you have, too. The inscription below a statue at the athletic venue at the United States Military Academy at West Point reads, "Upon the fields of friendly strife are sown the seeds that upon other fields on another day will bear the fruits of victory" (General Douglas MacArthur). Learning from competitive athletics happens every day.

Getting Buy-In

Everyone needs to get on board. The best way to do this is to involve them from the very beginning. As soon as the organization thinks they want to try VIRO, tell everyone about it. It can't hurt, and the most enthusiastic people will probably be those that count the most...the workers, the people in the trenches. They've known all along that management was confused about their safety. And guess what? They've had most of the answers most of the time. They've always felt that there was a lot of hypocrisy when it came to their safety. VIRO will either prove them right or wrong. It will shine a spotlight on any organizational hypocrisy with regard to safety.

IMPLEMENTATION STRATEGIES

As I have mentioned earlier, the best model for any organization is the family unit, whether it be a traditional unit or some variation. People living together form the basic unit for society. Other organizations, like clubs, church units, business units, or larger organizations, are just extended families, and what will work in the micro units will usually apply in the macro units.

Having said that, perhaps the best way to illustrate how to introduce the concept of VIRO and then the specific values is the same way you would do it in your family. A good model of how that can be done is provided in the book *Teaching Your Children Values,* by Linda and Richard Eyre. The Eyres suggest that the toughest question is not *what,* but *how.* That is what this section is all about. For VIRO, the *big* question is not *who*, but *how.*

The Eyres' Introduction addresses the questions: Why? When? Where? Who? What? and How? Parallels with VIRO are natural.

Why?

When it comes to children, the Eyres suggest that teaching values is the best way to insure their happiness. Corporate happiness is profitability. That is why there should be a foundation of safety values.

When?

My answer is the same as theirs: "NOW and ALWAYS." This is a large and complex job. So, the sooner you start, the quicker you will harvest the benefits. Can you afford to wait and allow someone else to achieve a business advantage? A trip around the world starts with the first step.

Who?/Where?

Their answer is in the family, and my answer for enhanced loss prevention is in the work groups and work place. Who needs to be involved? As I have said before and will continue to say: "All of the above." There isn't anyone who should be a spectator in the VIRO process.

What?

Here they point out that values are not universal; they must be personalized. Their interesting definitions apply as well to safety values. They describe, *"A true and universally acceptable 'value' is one that produces behavior that is beneficial both to the practitioner and to those on whom it is practiced.* " And, "It is something that *helps* or something that prevents *hurt.*" In addition they suggest that "A *value* is a quality distinguished by: (a) its ability to multiply and increase in our possession even as it is given away; and (b) the fact (even the law) that the more it is given to others, the more it will be returned by others and received by ourselves." "Values, then, are other-and-self-benefiting qualities that are given as they are gained and gained as they are given." They suggest that in today's world, too many people are concerned with *getting* rather than subscribing to a foundation of values focusing on *being and giving.* Just how much would basing conservation of your physical and human resources on a foundation of values enrich your organization, whatever its size, nature or mission?

How?

As suggested earlier, this is the *important* question. Recall that the keystone of VIRO is that values, beliefs, and attitudes intertwine to form cultures and determine behavior; and, beliefs/values/attitudes (i.e., culture) *can* be taught. How? By precept, by example, and by education and edification. These are the ways to sell VIRO and should clarify who should do what—obviously, the pieces need to be properly placed, and each individuals' positions and fortés should be used to full advantage.

Below I adapt the Evres' three approaches for use on our project of implementing VIRO.

APPROACHES FOR IMPLEMENTING VIRO

By Precept

In an extended family or group of some size, documenting the beliefs and values ensures that all will be aware of the principles aspired to by the group and the standard of conduct or action expected. This *does not* mean a list of values developed by the public affairs department, approved by the CEO, and used to paper the walls of most of the rooms in most of the buildings owned and operated by the enterprise.

The wall papering exercise that has been applied to mission statements and quality statements cheapens the human aspect of what needs to happen. The posted bills quickly become part of the landscape and are ignored by everyone who matters, with the possible exception of an occasional guest. This is akin to the Pharisees praying in public—it's all show.

The precepts should be developed by a broad cross section of the organization with guidance and parameters set by the leadership. It might not be a bad idea for everyone to have a copy of them at their desk or work station, but companies should avoid papering the walls with them. A few copies in strategic places would be okay, but they shouldn't frame hundreds and then look for places to hang them.

The important thing is for the members of the organization to relate to and understand the values and feel comfortable with them. This can only be done through time, with continuing effort and involvement.

By Example

Example *is* the great teacher. Everyone in leadership positions should demonstrate the beliefs and values of the organization *all* the time. The most subtle and sincere forms of expressing the values will be the most effective. Here are a few illustrations (i.e., examples) of what this means. They will be more effective than an extensive treatise on demonstrating how to set a good example. Deep down in their hearts, people know how to do it. Would a child be more impressed with the spirituality of their parent if they said grace before meals or if they happened to observe them in private, sincere personal prayer?

The Eyres, too, see example as the best teacher. Unfortunately, it is not something that most organizations recognize or do very well. Here is the organizations' chance to build skills in that area. The Eyres suggest teaching techniques such as verbal games and "scenarios," concept discussions, and praise with reinforcement. They suggest that "real change comes through catching children doing something good and then praising and reinforcing the behavior." Does this sound familiar to those of you who are devotees of behavior-based safety?

DO:

- Refer to meaningful measures of safety performance in discussions with colleagues and subordinates.
- Interject the personal aspects of accident prevention into discussions of protecting people, property, and the environment.
- Integrate loss prevention into every aspect of business: training, capital projects, new products, purchasing—the list is endless.

- Capitalize on every opportunity to recognize the benefits of conserving organizational resources. Retell stories of successes. Recognize and record them.

DON'T:

- Issue trite memos that look like they were designed to make someone happy—even OSHA inspectors.
- Champion programs that have failed in the past and have little support at the grass roots.
- Denigrate or make disparaging remarks about any aspect of safety or depreciate the contributions of any practitioner or participant. To do so is deadly.
- Issue any statement that you don't intend to back up with every fiber of your being.

By now, you probably get the point. The idea here is sincerity. It is no less essential here than in any other aspect of life that really matters.

By Education and Edification

Next to example, old-fashioned instruction is most effective when delivered by stories, games, and discussions. There are almost endless illustrations of how these techniques can be used by someone who is creative and sincere about their message. This section will offer some illustrations.

By Stories

The best stories include analogies, similes, metaphors, parables, allegories, anecdotes, and personal experiences. These are things to which people can relate and understand. The stories could be about near-misses or situations where someone avoided exposures by shrewd planning or application of the company values. Those real life situations can reveal side benefits associated with diligent adherence to agreed upon principles. Leaders should look for these stories like they look for new products or better manufacturing practices. They may be harder to find but will be less expensive and often provide more long-term benefits.

By Games

This doesn't mean pick-up basketball or racquetball during lunch time. It means activities that teach important concepts within the framework of having some fun and perhaps even lighthearted competition. Look at the survival programs that have become so popular in some circles for executives trying to do team building. Games don't have to be expensive or restricted to executives. The Factory Mutual organization used to have board games to teach people how to handle a fire system impairment and an emergency. The latter was called "Fire at Mansfield." I used them with plant managers and staff. They had fun and learned a lot about the adequacy of their programs. When they made a mistake, they burned down the Mansfield factory instead of their own. Their mistakes prompted corrective measures; years later, they remembered the exercise.

By Discussions

These can be held as an integral part of staff meetings. The beliefs and values will provide an endless reservoir of meaty subjects to discuss as appropriate.

Before concluding this chapter, it is important to review some potential pitfalls with suggestions on how to avoid them.

AVOIDING PITFALLS

Don't Allow the Concept to Get too Complicated or Violate the Rule of Simplicity

The basic concept here is simple. We need to accept the thesis that the foundation of every successful effort is a belief in the concepts that are sure to correlate with success. We then need to be convinced that beliefs and values can be taught...and that we can teach them. *No more is needed to make VIRO work.* The saying, "The devil is in the details," certainly applies here. But, if we lose the primary tenets, either in the beginning or along the way due to dilution or pollution, we will get lost on our trip. The way to avoid this is to *rivet on the basic concept* right from the beginning and not allow anyone to dilute or pollute it.

PEANUTS   By Charles M. Schulz

Make Sure the Organization Is Ready

Only the most sophisticated organizations will be able to implement VIRO. The organization will need balance between decentralization and centralized decision making. They will have to have a history of success in implementing similar efforts and have a safety process that has been generally successful. The only exception to these prerequisites would be a situation where a new management team has taken over, is in the process of a company make-over, and sees this as a model that can serve as a beacon for their efforts. If the organization is not ready and wants to do VIRO, they need to get ready. Getting ready requires a determination of where they are and then what they need to do to prepare. A good parallel (discussed earlier in this chapter) is what is needed for an organization to prepare to move to self-directed work groups. Many of those efforts failed due to inadequate preparation.

Don't Assume that You Can Start at One Part of the Organization and Migrate to Other Parts

Installing the process must start with convincing the decision makers that it should be done. At that point, the process must move ahead in lock step at every level of the organization. This is not a top down, bottom up, or a middle out effort. Everyone must get on board at the same time. If a company is not prepared to orchestrate a comprehensive introduction and follow up, *they shouldn't get started*. In a family where everyone belongs to a different church, not only will it be hard to worship together on Sunday, but there will be considerable diversity in what family members do each day based on the difference in beliefs. But, converting everyone to the same religious beliefs will require

the support of the family leader. In a more complex organization, that is the CEO, Executive Committee, Managing Director, or person in charge.

Grasp the Full Vision of the Power of the Concept and All the Benefits that Can Be Achieved

This is a group changing exercise. It will be hard to accomplish and will require substantial energy and commitment. That energy can come from understanding the enormous benefits that VIRO could bring to an organization. Success may not come in any other way. You shouldn't get started until there is not just an acceptance, but an infectious enthusiasm.

Be Willing to Pay the Price for Success

Benefits are invariably proportional to the efforts or resources expended. The rewards for laying a new foundation for how an organization protects and optimizes its resources are almost incalculable. There may not be anything a group could do that would have greater and more long lasting benefits. Hence, loss prevention success based on values will come at a high price. But, the price will provide the greatest return on investment that any effort has or could provide. This is a life changing process that should only be undertaken by high rolling organizations or those who feel a need to reinvent themselves.

TACTICS, PLANS, AND TIMETABLES

Once a company knows where they are and has a vision of where they want to get to, the next step is developing a strategy for how VIRO can get them there. Then comes the easy part, the plan. Assignments and timetables follow. These are so much a part of how every organization functions and so dependent upon the nature, culture, and needs of that group that it would be useless to explore this area in any depth here.

But, before concluding, let's reflect back on *Teaching Your Children Values*. In the preface, the Eyres write, "Your value system may be very similar to—or it may differ from—ours. The important thing, we feel, is that parents consciously develop their own set of family values and work at teaching those values to their children." This is one of the main messages in this chapter—if you substitute leaders for parents, organization or business for family, and employees for children.

The Eyres write, "If children start from a values vacuum— with none taught, none learned—they will float at the mercy of circumstance and situation, and their lives will never be their own." If safety is not based on a foundation of values, the entire organization is floating and has no firm reference point. Is it any different? In closing the preface to their book, they suggest that the debate over what family values are is silly and mostly a self-serving political debate. They suggest that parents already know what the values are and then suggest twelve that can serve as a reference point. I've suggested ten safety values. Organizations with successful safety efforts hold most, if not all, of these values, even if they don't call them values or have them in writing. The next chapter is all about documenting and quantifying those values.

18

MEASURING RESULTS

"Without measure, medicine will become poison."

. . . .

"If you can't measure, you can't manage." Peter Drucker

Right from the beginning of developing VIRO, I have felt that if I couldn't find a way to measure its performance, it would never progress beyond a dream, or at best, a narrowly held idea. Early in the process, I became convinced that measurement was feasible. Then, in a "rush" of inspiration one day, I became aware of how it could be done. It was an exciting time for me, since I was sure that without it, all my pondering would come to naught.

What I am going to suggest is not a finished or a one-size-fits-all product. For those who are lazy or are always looking for something for nothing, you should be warned. This is not a quick fix. If you routinely buy lottery tickets and have never won anything but keep trying, you may not be captivated by this chapter and its rationale.

These ideas target those who like to shape their own future, but are open to someone helping by sharing some ideas; who like new approaches and see adversity and complexity as lubricants for growth instead of catalysts for discouragement; who know there are better ways to avoid bad outcomes that are as yet undiscovered; who love to be at the cutting edge; and who get a charge out of improvement.

If you are still interested, begin reading about the issues that are immutable and those that will require at least some adaptation, maintenance, and/or mid-course correction.

Your organization needs to hold on to several fundamentals:

- An abiding belief that you can measure anything if you are creative enough.
- A belief that if loss prevention is right—and it obviously is—it is only a matter of time before you can build solid inertia for what needs to be done.
- An ability to predict behaviors and outcomes if beliefs and values are known.
- A willingness to share your beliefs and values if they are not threatened by the process.
- An understanding that the relationship between what you believe and what you do is infinitely more important than number results.
- A belief that VIRO has the *potential* to enrich more than conventional loss prevention. And, values-driven safety must provide unique benefits if it is ever to be widely practiced.

Issues that require further development and individualization are:

- Tailoring values for specific organizations based on history and complex demographics.
- Creating maturity definitions that are as smooth and understandable as possible.
- The importance and resultant weighting of the values to predict outcomes with the greatest precision.
- Administering the maturity grids to the organization and then learning how to use/interpret the results.
- Getting started by selling the concept and getting people on board.
- Correlating with existing measurement systems.
- Following up on measures to ensure that nothing good in the current process is lost.
- Establishing techniques to phase out activities that are found to be inconsistent with established beliefs and values.

The benefits of VIRO and this measurement system are legion and should become legend. They include:

1. People become conscious of what values the organization holds.
2. Culture development occurs naturally.
3. Natural selection takes over in hiring and promoting.
4. Everything important is out on the table. Hidden agendas are less likely.
5. People don't have to think or ask "Why?" The reason becomes apparent.
6. The measurement is predictive instead of reactive. It inspires confidence and security.

7. VIRO sends the correct message—this is a process, not an event.
8. VIRO provides never ending and meaningful fodder for management consideration and avoids safety meetings that often seem to be in search of an agenda.
9. VIRO models techniques that will enrich the organization in ways that are hard to imagine.
10. Interviews that will inevitably develop from the use of the maturity grids discussed below probe to the roots of dysfunction or wrong outcomes.

DOING IT

The measurement system for VIRO is based on maturity grids and surveys. These were introduced in Chapter 15. Their relevance relies on the assumption that organizations have cultures and that those cultures are based on that organization's beliefs, the consequent values, and the behavior naturally flowing from them. In other words, individuals in the organization express their values in their daily behaviors. And, others in the organization have the capacity to recognize and evaluate those behaviors. Lastly, it is assumed that the most accurate appraisals will come from subordinates when they look at their leaders.

That final assumption probably calls for some explanation. Bosses usually see their subordinates infrequently, and then the underlings are on stage and on their best behavior. They are actually trying to create an impression and are covering up any blemishes they have. Distortion results. When colleagues look at each other, jealousy, competition, and other fog factors may blur the picture. Subordinates usually see their leaders most clearly. They not only see them directly but through the behaviors and sayings of other members of the organization.

If the boss is good, the subordinates will develop a loyalty that may approach worship. If the boss is bad, those who work under them will ignore much of what they do and, in some cases, undermine their efforts—hoping for failure and then reveling in it. Good bosses are loyal to their subordinates and don't ask them to do anything they wouldn't do themselves. They accept the blame for failures and give credit for accomplishments. Bad bosses shed blame and take all the credit. They use and abuse their underlings. The bottom line is that people will evaluate their leaders with great candor and accuracy, particularly if they can remain anonymous. This process allows for that anonymity. There is no need to know who said what about whom as VIRO is not attempting to change individuals at this point. Rather, it is trying to change the whole organism. It will eventually need to change individuals but it doesn't need to do that individually. It can be done as part of the process, and it will happen as a natural result of what is being done. Most people will see that as they understand the entire theory and process.

Let's get to the heart of measurement. I will explain my vision of implementation details and options later. What is to follow is *one perception* of levels of maturity based on *my model* of sample values. Hopefully, they will provide a stimulus for others to devise their own beliefs, values, and maturity descriptors, but no specific set of safety values should be seen as universal. Every set should be unique to each organization, based on their history, needs, and goals. I have chosen to use six levels of maturity. Anywhere from three to ten are feasible, depending upon how precise you desire to get. Later I will also offer a model form on which the following information can be displayed and which can serve as the primary tool for implementation.

Here we go:

Value # 1 Do It for the Right Reasons

The Ideal:

You *believe* that your success is tied to the quality and health of your employees or members of your organization.

You *value* your human resource above all your other resources.

Maturity Level:

0—Safety is driven by regulation, management directives, and the cost of accidents.

1—Concern for people is occasionally mentioned, but the organization does not back it up with actions, and employees don't believe it.

2—Cynicism about the organization's interest in people's safety exists, but there are flashes of real concern.

3—Concern for people is balanced with regulatory compliance and the cost of injuries, but is seen as an independent subject.

4—Concern for people is a major safety driver, but it is not in *real* harmony with other organization goals.

5—Safety is driven by a sincere and often verbalized concern for employees and organization vitality; and the two are balanced in perfect harmony, just as safety, productivity, and quality are balanced in perfect harmony.

Value # 2 See It as Part of the Whole

The Ideal:

You *believe* that efficiency and an integrated approach to loss prevention are synonymous and essential to business success and optimization.

You *value* synergy and an integrated approach to the way you manage and eschew the "Humpty Dumpty School of Management."

Maturity Level:

0—Safety is handled independently of the rest of the management process.

1—At times, safety is considered at staff meetings and in appraisals, but this is infrequent and generally viewed independently of other management systems.

2—There have been efforts to integrate safety into the management process, but they are token efforts and have mostly failed.

3—Safety has parity with other staff functions and management has a vision of how safety should be managed.

4—A plan has been laid to fully integrate loss prevention into the management process, and independent discussions of safety are becoming infrequent.

5—There are no safety committees, safety talks, safety inspections, safety incentive programs, etc.; and loss prevention is an integral part of the management psyche and accepted as essential to business success and vitality.

Value # 3 Recognize There Is No End

The Ideal:

You *believe* that hard work is rewarded, that there is no short cut to safety excellence, and that it requires constant and never-ending pursuit.

You *value* persistence and tenacity when it comes to protecting assets.

Maturity Level:

0—Safety is managed by a series of quick fixes, and reaction is far more prevalent than developing prevention strategies.
1—Occasionally, efforts are made to determine root causes and some preventive efforts that don't have immediate payoffs are attempted.
2—Awareness is developing that loss prevention is hard work and that long-term success is not secured with incentive programs and punishing offenders.
3—Management is initiating efforts that are self-perpetuating and is demonstrating their long term commitment to protecting their physical and human resources.
4—Critical behaviors and conditions are being defined and systems developed to measure and manage them. Management does many things to indicate a long-term commitment to avoiding undesired outcomes.
5—Everyone in the organization recognizes that the pursuit of excellence in safety is like the Shewart Cycle, never ending.

Value # 4 First It Is a People Business; Things Are a Distant Second

The Ideal:

You *believe* that the actions of people are critical to an accident free environment and that facilities are nice, but rarely predict good outcomes in and of themselves.

You *value* the power of people and recognize that they can overcome almost any obstacle, certainly a weak procedure or substandard facility.

Maturity Level:

0—Safety is keynoted by inspection of physical facilities, regulatory compliance, and procedures manuals. The recognition of the human element is largely confined to incentive and training programs of uncertain value/success.

1—Some successful training is done, and employees have some sense of participation in loss prevention efforts.

2—Employees are beginning to feel ownership of some aspects of protecting themselves. Management is focusing more on behavior than conditions.

3—The focus is clearly on employee involvement and behavior-based safety; regulatory compliance and inspection of physical defects is seen as proceeding naturally from correct attitudes rather than as an end in itself.

4—A well-developed and internalized process exists to conserve physical and human resources based on measuring and monitoring people's actions that predict outcomes.

5—The organization knows what values, as well as actions, predict desired outcomes. They focus on attitudes/culture/values that predict behaviors consistent with accident/incident-free performance, know where they stand at all times, and can correlate that with other more traditional performance measurements.

Value # 5 Put the Right Person in Charge

The Ideal:

You *believe* that safety results are directly related to the expertise and skills of the people put in charge of the subject.

You *value* conservation of your physical and human resources enough to insure that they are only entrusted to the very best people you can find and place in your organization.

Maturity Level:

0—No emphasis is placed on who heads up the safety process, and the practitioners are rarely qualified or among the most competent people in the company.

1—Token efforts to staff for safety are made, but they are hit or miss and results are not much above start-up.

2—The reporting relationships say that safety is not as important as many other management functions. Safety jobs are most vulnerable when there are cut-backs; incumbents are rarely well-qualified or well-respected.

3—Reporting relationships do not detract from the safety process. Safety professionals are generally up to the tasks they are called upon to accomplish.

4—The loss prevention function is responsible to an effective executive. Safety people are generally competent and sufficient in numbers to accomplish needful tasks.

5—The function reports to an influential executive who can and does go to bat for safety. Only the best people are placed in all *necessary* safety positions. They either come well-qualified or are systematically developed with "world-class" as the stated goal.

Value # 6 Use a Yardstick Everyone Can Read

The Ideal:

You *believe* people want to be measured (fairly) and that accurate measurement is essential for improvement.

You *value* helping everyone access accurate, fair, and understandable safety performance measurement.

Maturity Level:

0—Safety performance measurement is almost universally misunderstood and is rarely referenced in any serious business arena.

1—Safety measurement is mentioned...but with little conviction of its correctness or worth, doubtful understanding of its real meaning, and without substantive responses.

2—Some people in the organization have a working knowledge of their safety measurement systems and how they relate to each other. Awareness is growing and at times elicits responses.

3—The organization has generally recognized and publicized safety measurement systems, although they're primarily retrospective. They generally elicit responses, although they are only infrequently targeted with predictable results.

4—Measurement of safety is clear and generally understood. It helps set agendas and is moving toward more prediction than reaction.

5—Safety measurement is prospective, positive, and credible, correlating with everything important in the organization. It is an effective tool, recognized as such by everyone.

Value # 7 Sell Benefits...and There Are Many

The Ideal:

You *believe* that the safety process is part of your solution and not one of your problems, and that it has many spill-over benefits to other corporate interests/areas.

So, you *value* the benefit of recognition and reinforcement of that process and talk about your feelings frequently and eloquently.

Maturity Level:

0—Everyone talks about the burdens of safety. Benefits are rarely discussed and not really appreciated.

1—Occasionally someone points out the virtues of loss prevention, but it is the exception rather than the rule.

2—Safety is sold in company publications and at times at meetings, but it is superficial and separate. The response is casual and unconcerned.

3—The benefits of loss prevention are routinely touted and with some conviction. It is part of the management mantra, but is not fully integrated and seems to come in waves.

4—Sophisticated selling of the benefits of safety benefits is evident. Virtually all employees "buy-in." It is socially and politically correct to talk about the benefits of working safely.

5—All the benefits of safety are known by everyone, including the sure knowledge that practicing the principles can enhance every facet of business life. Talking safety benefits is inherent in the organization culture.

Value # 8 Never Settle for Second Best

The Ideal:

You *believe* that safety is synonymous with business success.

You *value* every facet of loss prevention and never subordinate it to other concerns.

Maturity Level:

0—Safety is the last item to be given budget consideration and the first item to be cut during hard times.

1—At times loss prevention has parity with other considerations, but that is clearly the exception rather than the rule.

2—Safety posters and placards are evident, as is a lot of *talk* about "safety first." When it comes to spending time and money, the phrase "you talk the talk but don't walk the walk" immediately comes to mind.

3—In most cases, safety is given equal time and support with other issues. When push comes to shove, there are obviously higher priorities.

4—Loss prevention is considered to be an essential element of business success. The specialists who facilitate safety enjoy parity with all other disciplines in the company.

5—Safety is not viewed as an overhead item. It is seen as a profit center and a way to gain a competitive edge. It is never subordinated to other concerns. The people who guide the safety process are considered valuable organization assets.

Value # 9 Be Guided by Logic, Not Emotion

The Ideal:

You *believe* that safety should not—must not—yield to emotion.

A high *value* is placed on rational, well-supported systems that predict desired outcomes.

Maturity Level:

0—Safety is always an emotional issue. The only way to get action is to get excited.

1—Loss prevention is said to be part of the management process but is driven differently. Action is invariably prompted by accidents, employee pressure, and regulatory doom and gloom.

2—Management knows that they should act rather than react when it comes to safety and is making efforts; old habits are dying hard.

3—Loss prevention is driven by a process that most people are confident works. Appeals to emotion still derail the process more often than it should.

4—Emotion rarely affects loss prevention decision making. But, occasionally pressure will prevail over prescience.

5—Management refuses to react to anomalies. They have complete confidence in their process and their ability to work it and refine it.

Value # 10 Empower Others Rather than Seeking after Support

The Ideal:

You *believe* that effective loss prevention is rooted in proper placement of responsibility.

Value is placed on the safety professional as an enabler and facilitator...nothing more or less.

Maturity Level:

0—If the person charged with safety responsibility doesn't do it, it doesn't get done. The safety resource person reacts accordingly.

1—The safety professional plays a dominant role, is somewhat insecure, and gives up tasks reluctantly.

2—Empowerment is starting, but without the safety professional, the process falls flat on its face.

3—The organization is beginning to understand the role of the staff support person. Management and workers are reluctantly taking responsibility, with frequent reversions to dependency, which is accommodated.

4—The safety professional is beginning to work himself or herself out of work. He or she is secure, has a strategy of empowerment, and is working it.

5—The organization has shed its dependency on the safety professional. They know how to use the function to optimize performance. The safety professional is comfortable and secure with the maturity of the organization.

USING THE MATURITY DESCRIPTOR-PREDICTORS

This maturity material or your customized equivalent should be on a one sheet matrix using both sides of one piece of paper. It should be supported by a simple manual or one-page, two-sided handout for reference that could be given to the evaluator. They could even receive basic training on how to do the evaluation that would be more of a question and answer session, as this process should be as simple as it is profound. A sample of those two tools is provided at the end of this chapter. The grid is labeled in the first figure at the end of this chapter. The sample instructions are labeled.

WEIGHTING AND USING THE RESULTS

Reducing complex subjective evaluations to numbers is something almost everyone likes to do. Engineers have a particular affinity for attempting precise evaluations of imprecise subjects. Unfortunately, not many subjective analyses lend themselves to such exacting interpretation. Luckily for me, educated as a chemical engineer, I think this one does.

The explanation starts with the basic tenet that the values as a composite can be designed to predict success or failure, accidents or an accident-free environment, and good or bad outcomes. Some may be more important than others. The suggested system of weighting can accommodate that.

Here is how it works. When numbers are used, people seem to be most familiar with a decimal system, and most comfortable with a base of one hundred. That can be done here. If you just add up the numbers from the ten value evaluation and multiply by two, presto: the maximum score is one hundred. Now, suppose the customized list of values is more or less than ten. All you do is multiply the final score by ten divided by the number of values. If a value is thought to be twice as important as another, just use a multiplier to weight the importance of the values. You just need to make sure that each increase in one multiplier is balanced with a reduction in another or apply some other technique to maintain the one hundred maximum score.

The result is a single number that predicts process safety performance, as long as the premises are correct, the correct values have been chosen, appropriate maturity descriptors have been applied, and an adequate and representative sample of employee evaluators has been used. Those conditions must be met, but they are reasonable and achievable. As the process is refined, predictability will be enhanced. In addition, we will learn, have fun, derive wonderful side benefits, and harvest a process that can be applied to other areas of interest. In the example given at the end of the chapter, the process is illustrated using a hypothetical company.

ADMINISTRATION

It is suggested that each organization uses a minimum of twenty observers or evaluators. They should be a cross section of employees with a bias toward the areas that have the greatest hazards or where the incidence of mishaps are the greatest. For consistency, the evaluators should remain the same for at least some finite time. The person who administers the activity should be independent and observe total confidentiality. There is no need to know who responded in what way. Discarding the results at the top and bottom ends of the evaluation may improve generalizability, although that may not be necessary.

If the organization is heterogeneous, evaluate the parts as well as the composite. This will begin to project who may be headed for a collision. The strong can help the weak on a basis that is more likely to engender receptivity than the traditional measures. The *apparent* good performers are always suspect, and the weak are more likely to see the so-called strong as better at keeping records than better at loss prevention. This VIRO process should place people on an even playing field with which they are all comfortable. The results can be used just as a thermometer would be used, with a single easy-to-understand dimension.

REFLECTIONS ON SELLING THE CONCEPT

All members of an organization will need the desire to embark on the use of VIRO or success is unlikely. VIRO strikes a respondent chord with almost everyone who hears it. During formal and informal discussions about a value-inspired approach to managing safety with a cross section of people, no one has ever responded negatively. The concept sells itself, *if people are prepared to listen*. If they are not, it may just be bad timing. It has been said that "One can only see rightly with the heart; what is essential is invisible to the eye." This concept appeals to the heart as well as the eye, maybe more so. It feeds off the concept of the importance of what makes humans unique in the kingdom of living beings. It is suggested that if you have charm, nothing else matters. And, if you don't have charm, nothing else matters. This concept has charm.

CORRELATION WITH EXISTING MEASUREMENT SYSTEMS

At one point, Chesebrough-Ponds correlated audit scores with injury incidence rates and workers' compensation scores. They never were able to achieve a precise relationship due to the variables in different states and even different countries, not to mention different businesses. However, within brackets of twenty percent or so, there was almost amazing consistency. They never had someone ranking in the top fifth in one category who was in the bottom fifth in any other, or even close to it. For the most part, these are the only three crude measurement systems available to safety practitioners. Here is a new one that should be considered the ultimate. But, others can still be used until they are found to be not of much use, if that time ever comes. I think it will.

If you have already compared measures currently in use, you have a head start. If not, using VIRO will provide the impetus to do it. The goal is to ultimately use the organizational culture to predict safety performance. Comparing VIRO results to audits, workers' compensation costs, and incidence rates by using graphs and displacing the lines or bars for time disparities will be interesting and enlightening. If these premises are correct, over time, your ability to predict workers' compensation costs, incidence rates, and audit results will be enhanced. Eventually, you will be able to foretell them with precision.

Since VIRO is based on a deep foundation of psychology, theology, and years of loss prevention experience, using and measuring it should refine and give greater meaning to the other ways safety is measured, as flawed as they are. This fourth method of measuring should help to focus and refine the other measurement techniques.

EXAMPLE

Introduction

In order to illustrate the maturity grid concept, I'm going to offer an example. I'll describe an organization and then suggest hypothetical results from evaluations using the maturity grid. Then, I'll use a weighting we will assume the company devised during focus sessions with representative employees. The result will be a composite maturity and a single numerical rating of the hospitality of the company culture (based on values held by employees at different levels in different disciplines and parts of the organization) to accident-free performance.

Company Description

This fictitious company is over one hundred years old and was family owned until going public over thirty years ago. The family still owns significant stock, but has no representatives in senior management or on the board and exercise very little influence. The core business was basic chemicals, but the company now is involved in several other sectors. The founder was very paternalistic, but time and corporate events have eroded much of the culture he and his son installed from the founding until over fifty years ago when professional managers took over.

The company is a Fortune 200 Company, with thirty thousand employees in twenty-five countries. Business units hold to some very general philosophies but operate independently, with very little cross-fertilization. Movement from one business unit to another is unusual. Disciplines are sophisticated but only interface to the extent necessary for routine business purposes.

Profitability has been excellent for the entire history of the company, and stock value is based, in part, on consistent and solid growth. Research is important in several business sectors, and the company supports it based on the need for cutting-edge technology that is deeply imbedded in company culture. The company was one of the first in American industry to have a safety function and full-time professionals. Their performance has been good, but has never been described as "world-class" or even best in class.

They have measured performance using traditional methods with a strong focus on incidence rates and lost time cases. Few employees are privy to workers' compensation costs, and they are only rarely discussed at management meetings. The organization has had a safety audit, but is disenchanted with it and is doing some benchmarking to update it. The method of involving employees in safety and publicizing it is very traditional and follows familiar models developed and advocated by OSHA, the insurance industry, the National Safety Council, and trade associations.

The company has tried self-directed work groups in certain sectors with mixed results. A few have been very successful; some have floundered; a couple failed. Leadership would like to widen the practice but are apprehensive and don't seem to be well-aware of the ingredients needed for success.

Due to recent downsizing and reengineering, unfamiliar to most employees, morale is lower than at any time in the company's history, and employees are apprehensive about job security. The corporate safety department has been depleted through attrition and has assigned people to operating units. The leader has been reassigned from the corporate counsel to the risk manager in the finance department.

Safety performance has begun to show some deterioration, particularly in labor intensive divisions.

Process

Forty employees across the company, representing five businesses and at different levels and with varying responsibilities, are provided with tools similar to those at the end of this chapter: maturity grids and instructions on how to use them. These were developed internally over several months by teams representing all levels of management and non-exempt employees. For now, assume they used this grid and that the results were consistent and close enough to one of the six maturity levels to round to a whole number. That will make the calculation easier to understand.

Results

The following are the results of the calculation:

1. Do It for the Right Reasons	2-Novice
2. See It as Part of the Whole	1-Entry Level
3. Recognize There Is No End	2-Novice
4. First, It Is a People Business; Things Are a Distant Second.	1-Entry Level
5. Put the Right Person in Charge	1-Entry Level
6. Use a Yardstick Everyone Can Read	2-Novice
7. Sell Benefits...and They Are Many	3-Mediocrity
8. Never Settle for Second Best	3-Mediocrity
9. Be Guided by Logic, Not Emotion	2-Novice
10. Empower Others Rather than Seeking after Support..	2-Novice

Maturity Score Computation

Value #	Score x 2 (Raw Score)	Multiplier+	Adjusted Score*
1	4	1.5	6
2	2	0.5	1
3	4	0.75	3
4	2	1.5	3
5	2	0.5	1
6	4	0.75	2
7	6	1.5	9
8	6	0.5	3
9	4	1.0	4
10	4	1.5	6
Total	38	10	38

* A multiplier was used to achieve a maximum score of 100.
+Factor indicating perceived importance. Total must always equal 10.
Note: It was coincidental that the adjusted score equaled the Raw Score. If important items had high scores, the adjusted score would exceed the raw score or vice versa.

Analysis

Management is surprised by how low their score is but accepts the results and commits themselves to learning from them. They recognize that they are perceived as being best at what they have tried hardest to do—sell safety. With performance deteriorating, they are debating whether the selling is as important as they thought it was. They have the feeling that their efforts are based on a weak foundation. They conclude that their initial weighting of the importance of values was probably out of balance and intend to take a fresh look at that.

The process has management and employees feeling that their safety efforts have started to lag behind other facets of company effort. They now feel that safety should be a leading edge, rather than trailing other functional areas. They can see the need for synergy and recognize that isolating safety efforts has weakened their process and contributed to deteriorating results. They see this exercise as an early warning and resolve to develop a strategy to improve their safety culture. When they put the whole maturity grid on a single page and highlighted the box in which their survey placed them, it triggered an animated discussion. "Was that the real rating or just the perception of employees?" "Have the generally healthy labor relations in the company served to make a weak safety effort produce better results than they should have had?" "Are results in steep decline?" No unanimous answers were immediately forthcoming. But there was agreement that management needed to probe and find the answers.

Management feels that values 2 and 5 need to be weighted more heavily and has resolved to invigorate the safety staffing, change reporting relationships, and place more emphasis on integrating safety with other aspects of business management. The exercise has illuminated their understanding of the loss prevention process and pointed the way to improvement before the situation became seriously degenerate.

Discussion

This exercise is obviously very superficial and conjectural. But hopefully, it gives you a *glimpse* of the utility of managing safety on a foundation of values that can help shape the culture of an organization. What can be learned from such an exercise is limited only by management's desire to probe the depths of what drives loss prevention efforts and potentially other aspects of their business.

Tools

There are two primary tools that can be used to survey subordinates: sample instructions for the employees to evaluate their leaders; and the maturity grid for the values.

GUIDANCE FOR SURVEY PARTICIPANT VALUE INSPIRED RESOURCE OPTIMIZATION: A Value-Driven Loss Prevention Process

Introduction

VIRO, Values-Inspired Resource Optimization, is a process to facilitate ever-improving loss prevention. The eventual objective is to minimize all losses, unwanted events, or undesired outcomes. VIRO is based on the causation chart in the following diagram, which describes how individual or group beliefs and values predict individual or group loss prevention results.

BELIEFS $\Rightarrow$ VALUES $\Rightarrow$ CULTURE $\Rightarrow$ ATTITUDES $\Rightarrow$ BEHAVIORS $\Rightarrow$ ACTIONS $\Rightarrow$ INCIDENTS (OUTCOMES)

We believe that if our organization ascribes to specific values that are designed to predict an accident and loss free environment, we can best assure our employees and owners that our business will minimize undesired outcomes.

We have determined what those values are for our organization [or substitute company name]. The determination was achieved after discussions with a cross section of employees at all levels, in many different parts of the organization. We are prepared to modify the values and how we define them as our culture evolves or as new information leads us to change our focus.

Measuring

We have determined that the best way to ascertain the maturity of our culture in relation to our desired value base is to look at the perceived values of a cross section of managers and supervisors in representative sectors of the company. We have developed a maturity grid that describes levels of maturity for our values to which we can refer as "predictors." The values are described in the attachment to the maturity grid that is provided. We call those explanations of what is meant by the target value at various stages of maturity as "descriptors." [Attach a description of the beliefs, values and maturity levels, such as listed in Chapter 18.]

Your Role As A Survey Participant

We ask you to check the maturity level that comes closest to describing the values you feel are held and modeled by the leader whose name is at the top of this document. In addition, there is an attached blank sheet on which you can offer input or make any observation about this process or your input.

Confidentiality

Do not put your name or the name of the person being evaluated on the forms. Our intention is not to evaluate *any* individuals. We seek only to gauge the status of organizational values linked to better safety performance. The return envelope is coded to insure that all surveys are returned. Once surveys are returned, the envelopes will be destroyed.

Questions

If you have any questions about any aspect of this process, feel free to call the envelope addressee. They are familiar with all aspects of this process and should be able to answer your questions.

Application of Results and Feedback

We believe that values can be taught. Constant efforts will be made to encourage all employees to embrace the values that are associated with an accident-free workplace. The results of these surveys will help us focus efforts and refine the methods used to teach these values. In addition, we believe it will predict for us what we can expect in the way of safety results in the near future, much as a barometer is used to warn of weather changes. As soon as the results have been assembled, you will be given a compilation on a priority basis. Explanations will follow, and you will notice numerous activities designed to increase our safety maturity.

VIRO MATURITY GRID (Part One)

Safety Value	0-Darkness Ignorance	1-Dawn Entry Level	2-Mid Morning Novice	3-Late Morning Mediocrity	4-Noon Excellence	5-Full Light Perfection
1. Do It for the Right Reasons	Safety driven by regulation, management directives, & cost of accidents.	Concern for people occasionally mentioned but not backed up with actions. Employees don't believe it.	Flashes of real concern for people occur, but cynicism is ever present.	Concern for people is balanced with compliance & injury costs but is seen as separate subject.	Concern for people is major safety driver but not in real harmony with other goals.	Sincere concern for employees drives safety and is in perfect harmony with other activities.
2. See It as Part of the Whole	Safety is handled independently of the rest of the management process.	Safety is occasionally considered at staff meetings and during appraisals.	Efforts have been made to integrate safety, but they have been token and often failed.	Safety has parity with other staff functions, and management has a vision of how it should be integrated.	A plan to totally integrate safety exists. Independent safety discussions are becoming less frequent.	Loss prevention is totally integrated and accepted as essential for business success.
3.Recognize There Is No End	Safety is managed by reaction & quick fixes.	Root cause determination & prevention efforts occur but are the exception.	Awareness that loss prevention is hard and ongoing is occasionally seen.	Management is initiating efforts that are self perpetuating. Evidence of a long term commitment to safety appear at times.	Critical behaviors and conditions are being defined & measured. A long term commitment is evident.	Everyone recognizes that safety excellence is never ending, like the Shewart Cycle.
4. First, It Is a People Business; Things Are a Distant Second	Safety is keynoted by inspection and compliance.	Employees have some feeling of involvement in the safety process but don't exhibit any sense of ownership.	Employees are beginning to have a sense of ownership for the safety process that goes beyond participation.	Focus is on employee involvement. Culture and attitudes are more important than physical defects	A well developed process for measuring and monitoring behaviors exists. The organization is moving toward deeper understanding.	The focus is on beliefs, values, and culture. It drives all other efforts and is correlated with all other measures.
5. Put the Right Person in Charge	No emphasis is placed on who supervises safety or on the qualifications of safety practitioners.	Token efforts are evident concerning safety staffing, but they have not been very effective.	Safety parity is a goal but is clearly not yet a reality. Everyone knows safety is often the first thing to be cut.	Reporting relationships are good and safety staff are generally well qualified and have adequate resources.	An effective top executive supervises safety. There are sufficient safety professionals and they are almost all highly qualified.	Only the best people are placed in safety positions. Their leader is an influential executive who goes to bat for safety.

VIRO MATURITY GRID (Part Two)

Safety Value	0-Darkness Ignorance	1-Dawn Entry Level	2-Mid Morning Novice	3-Late Morning Mediocrity	4-Noon Excellence	5-Full Light Perfection
6. Use a Yardstick Everyone Can Read	Safety performance measurement is not understood and is rarely discussed at serious business meetings.	Safety measurement is mentioned but with little conviction and without substantive responses.	Awareness of safety measurement is growing and at times elicits responses.	Safety measurement is good but largely retrospective. There are responses but rarely targeted with predictable results.	Measurement of safety is clear and generally understood. It is moving toward being more predictive than reactive.	Safety measurement is prospective, positive, and credible. It is an effective tool that correlates with all that matters in the organization.
7. Sell Benefits...And There Are Many	Everyone talks about the burdens of safety. Benefits are rarely discussed or appreciated.	Occasionally someone points out the virtues of safety, but it is the exception instead of the rule.	Safety is sold in company organs and at some meetings, but it is not integrated and response is casual.	The benefits of loss prevention are often touted—at times with conviction. But, it is not fully integrated and comes in waves.	The selling of safety is consistent and integrated. Employee involvement is "built-in." Talking safety is encouraged.	Talking safety benefits is inherent in the organization culture. It is fully integrated. Everyone is aware of the benefits.
8. Never Settle for Second Best	Safety is last when it comes to allocating funds and the first to be cut during times of austerity.	At times safety achieves parity with other considerations, but it is the exception, not the rule.	There is "safety first" talk and posters, but it is often not supported.	Safety normally has parity with other considerations but at times it is clear that there are higher priorities.	Loss prevention is considered essential to business success. Safety professionals are considered equals with other staff.	Safety is viewed as a profit center, not as overhead. Safety professionals are viewed as valuable organization assets.
9. Be Guided by Logic, Not Emotion	The only way to get action on a safety item is to get emotional.	Safety is said to be a part of the management process but is driven differently. Action usually stems from accidents or complaints.	Management is transitioning to act and not react and quest after root causes, but old habits persist.	Loss prevention is driven by process. Appeals to emotion still interfere more often than they should.	Emotion rarely affects safety decision making. But, occasionally pressure will prevail over prescience.	Management refuses to react to anomalies. They have complete confidence in their process and stick to it tenaciously.
10. Empower Others Rather than Seeking after Support	If the safety professional doesn't do it, it doesn't get done.	The safety professional plays a dominant role and delegates reluctantly.	Empowerment is starting, but in the absence of the safety advocate, the process loses momentum quickly.	The organization understands the correct role of the safety professional but frequently reverts to dependent behavior	The safety professional is working themselves out of a job. They are secure with their empowerment strategy and implementation	Dependency on the safety professional has been shed. The function is used to optimize performance. The safety professional likes it that way

19

BEYOND EXCUSES TO EXTENDED APPLICATIONS

"Inspiration for writing is mostly the application of the seat of the pants to the seat of the chair."

. . . .

"Success never needs an excuse."

. . . .

"Don't make excuses, make good."

I firmly believe—and hope I have persuaded progressive safety professionals and courageous COOs and CEOs—that the concepts suggested in the book have the potential to vault an organization to the top of any group with whom they compete. But the resistance will be enormous. While VIRO makes all the sense in the world, managing by values confronts many traditional organizational behavior patterns. This chapter exposes these patterns for what they are: emotional and nonsensical. After toppling these hang-ups, excuses, rationalizations, or whatever you want to call them, move on to explore VIRO's extended applications, where forward thinking professionals could ultimately take values-driven management approaches. You see, this thinking has the potential to change the world, after it is used to change the way safety and specific organizations are managed, particularly and initially those who operate for a profit.

Thirteen Reasons Not To Do This, or Why Managing Safety, Business, Your Personal Life, or the World on a Foundation of Values Won't Work and My Response

1. These concepts are "soft." They are not supported by science or, for that matter, any social research by an authoritative source. Admittedly, as demonstrated in Chapter 2, there is growing interest in values and related subjects. *That may be true, but given what is at stake, what have you got to lose by trying it? In the light of a decaying situation, who has a better idea? The populace will—if allowed—pass judgment on this concept quickly and decisively. Let's let them vote.*

2. No one has proven it will work. *Every good idea that has enriched mankind needs to be proven. This requires discussion and experimentation. You need a pilot project. You need people of courage and vision, like Bob Haas, the CEO of Levi Strauss, who is mentioned later. No one will ever know if this will work unless it is given a chance.*

3. You can't teach morality or values. *It has certainly been demonstrated that one can teach immorality or values that feature excuse making, blame placing, and abrogation of responsibility. Look around you. And it appears to have only taken one* *generation. The road must go in both directions, unless the bridges have been blown. Have they? Surely you can lead/teach in the other direction. Maybe one can't legislate morality, but people can live morally.*

4. People don't like to talk about things as personal as values and beliefs. *I don't believe this, based on my personal experience. People won't like to talk about values and beliefs only if the party with which they are conversing ridicules or treats them lightly. But if ingenuousness and a sincere intent to listen and respond define the setting, most people would be willing and perhaps delighted to talk about their values and beliefs.*

5. You can't measure improvement. *Hogwash! All you have to do is look around. A specific system has been provided in Chapter 18. Creative people could come up with other methods. If people believe in strong marriages and families and so value chastity, and if everyone or at least more people therefore start practicing fidelity and avoiding premarital sex, do you think our incidence of out-of-wedlock births will decrease? It's easy to fashion one hundred more questions that could likewise obviously measure the effects of values and changes in a group's values.*

6. Safety can't be a beacon. Nobody *really* **cares.** *This is the hardest reason to answer because it is essentially true...sad to say. Safety is non-threatening and bipartisan. Everyone is for it. The fact that nobody* **really** *cares could be an advantage. Resistance won't be as entrenched. Maybe it could start out as a flashlight and grow into a beacon.*

7. This is too complex and esoteric for the average employee to grasp. *They may not have read all the philosophy behind it,* *but: "Try it, you'll like it." The rank and file employees certainly will gravitate to this like a bear does to honey. They've probably been hoping management would champion the values they have held for years. They are dismayed by what they see around them. They're looking for leadership in this area. Working people are the heart and soul of this nation.*

8. Resistance to change is great. *It is difficult to dispute the reality of resistance to change, even positive change. As one who enjoys variety in food and life, I'm constantly amazed at how people resist even good change. But this isn't a great change. The concept of connection between goodness, values, and results is intuitive. People have just been confused and led astray. When light is shined on these things, people will hear a lot of "ah ha's" and "I know that's right." This is a change waiting to happen, particularly right now.*

9. No one has a program. *In a free enterprise system, products emerge to meet the market. If people or companies say they want this, everyone and their cousin will be jumping on the band wagon. I've been trying to figure out how I can corner the market on this if I'm right and it catches on. I've decided not to try. If it works, there will be plenty in it for everybody. Actually, this book includes a lot of practical information and tools.*

10. Everyone believes something different. Getting everyone to believe the same thing or have the same values will be impossible. *Changing what people believe won't be impossible but it will be hard. The harder it is, the stronger the process will be once the conversion is complete. We're not talking about building consensus on the nature and existence of God or some obscure religious doctrine. We're talking about unifying people on concepts that can be, and have been, proven in everyday life.*

We can get most people on board; the others just drown or go work somewhere else.

11. Not everyone buys into religion (believing). Those who resist, resist passionately. *This is true. But VIRO is not religion. It is non-religious, it has a social base, and it is built on concepts that are common to almost all religions, although you don't have to be a follower of any traditional religion to understand the wisdom of this process and practice it successfully.*

12. Predicting success using a values-directed approach to safety, much less all business or all the world problems, is very presumptuous. *Why should this succeed when so many others have failed? Where would the U.S. be if Washington, Lincoln, Edison, Martin Luther King, Columbus, Jonas Salk, and even Gorbachev could not see beyond their line of sight? All these men knew the importance of the perspiration that follows the inspiration. It certainly is hard to see yourself in the same picture that you see them, but if developed, this concept could be world changing, just as the things they did were.*

13. People would be wasting their resources to do this. *Developing a foundation of values upon which to base a safety program, a business, one's personal life, a country, or the world would not seem to be a waste of resources to me, based on all the out-right silly things on which people spend time and money. Have you ever seen those hot pink windshield wipers and license plate frames people buy for their cars? Have you ever watched an afternoon soap opera? Take a look at the latest fashions and their prices. How many man-hours were wasted watching the O.J. Simpson trial? How many dollars are spent on consultants who write reports that never get read, much less cause action? How many government studies study subjects about which nobody* *cares and will ever do anything about? Managing by values could produce the greatest return on investment that the investing person or organization could ever imagine.*

These hang-ups seem silly when measured against the magnitude of potential benefits directly to safety and easily extended into business, political, social, and personal dealings.

EXTENDED APPLICATIONS

In introducing my comments on the extended applications of values-based management, let me share a poem, by an anonymous author, that recently impacted me. Poetry often—for some inexplicable reason—has a way of getting to the soul to make a point in a way that prose cannot.

A Fence or an Ambulance

'Twas a dangerous cliff as they freely confessed,
Though to walk near its crest was so pleasant,
But over its terrible edge there had slipped
A duke, and full many a peasant.
So the people said something would have to be done
But their projects did not at all tally.
Some said "Put a fence 'round the edge of the cliff."
Some "An ambulance down in the valley."

But the cry for the ambulance carried the day
For it spread through the neighboring city.
A fence may be useful or not, it is true,
But each heart became brimful of pity,
For those who had slipped off that dangerous cliff,

And the dwellers on highway and alley
Gave pounds or gave pence, not to put up a fence,
But an ambulance down in the valley.

"For the cliff is all right, if you're careful," they said,
"And if folks even slip or are dropping
It isn't the slipping that hurts them so much,
As the shock down below when they're stopping."
So day after day, as these mishaps occurred,
Quick forth would their rescuers sally,
To pick up the victims who fell off the cliff
With their ambulance down in the valley.

Then an old sage remarked "It's a marvel to me
That people give far more attention
To repairing results than to stopping the cause
When they'd much better aim at prevention.
Let us stop at its source all the mischief!" he cried.
"Come neighbors and friends, let us rally!
If the cliff we would fence, we might almost dispense
With the ambulance down in the valley."

"Oh he's a fanatic!" the others rejoined
"Dispense with the ambulance? Never!
He'd dispense with all charity too, if he could.
No! No! We'll support them forever!
Aren't we picking up folks as fast as they fall?
And shall this man dictate to us? Shall he?
Why should people of sense stop to put up a fence,
When the ambulance works in the valley?"

But a sensible few, who were practical too
Will not put up with such nonsense much longer.
They believe that prevention is better than cure
And their party will ever grow stronger
Encourage them then, with your purse, voice, or pen,
And while other philanthropists dally
They will scorn all pretense, and put up a fence.
On the cliff that hangs over the valley.

Better guide the young, than reclaim them when old,
For the voice of true wisdom is calling
To rescue the fallen is good, but 'tis best
To prevent other people from falling.
Better close up the source of temptation and crime,
Than deliver from dungeon or galley.
Better put a strong fence 'round the top of the cliff,
Than an ambulance down in the valley.

Before moving on, let's analyze this poem.

In this community, emotion and the desire to do good overwhelmed logic. These people could see the fallen duke and all the peasants. They had to come to their aid. They just couldn't advance their thinking to see that the fence would obviate the need completely. The rescue work took on a life of its own. Surely those rescue squad people loved their shiny ambulance and the opportunity to serve. It was exciting and rewarding.

Finally, the old sage just couldn't contain himself anymore. He had enough experience, age, and security that he could put his reputation on the line. Then came the flack; resistance to change reared its ugly head. But eventually nonsense yielded to sense. Finally, the "sensible few" just wouldn't be shut up. "True wisdom" called and couldn't be silenced.

VIRO is the best approach for the "sensible few" safety professionals, and its "true wisdom" will *naturally* attract expanded applications. Next, you'll see an example of how enlightened businesses are already gravitating to a values-based approach. But before that is done, the stage will be set by drawing some additional parallels, as promised in Chapter 11. In addition to parallels between safety and quality, many of the statements in *Quality is Free* could apply to optimizing business and family relationships.

Let's start with seven that compare safety and quality management. They are:

The beauty of the application is that you don't have to be a quality/safety professional to apply it.

If something is easy to understand and makes sense, and yet isn't always done, there has to be a reason for not doing it. It is the challenge of safety/quality to determine that reason and lay it to rest.

You can get rich by preventing defects/accidents.

Quality/safety is free, but no one is ever going to know it if there isn't some sort of agreed-on system of measurement.

Quality/safety is free. But it is not a gift.

The real cost of safety/quality is "below the tip of the iceberg."

You need to get to the source to solve safety/quality problems. That is the employee who is nearest to the action.

Now, here are seven that don't have the words in them (safety or quality), but apply equally well to the management of safety and quality, and, in some cases, to many other areas.

Managers think they understand when they do not. (They oversimplify, not having adequate understanding.)

The field(s) have unique language that confuses managers.

They require a cultural revolution.

They are achievable, measurable, and profitable and can be installed if you have commitment, understanding, and are prepared for hard work.

They often "turn off" the management they were trying to entice.

They require unblinking dedication, patience, and time. At some stage, self-evaluations can be very effective.

Finally, here are seven statements that apply to the management of quality and safety, but clearly to many other activities important to business and the human family.

Prevention is not hard to do, it is just hard to sell.

Attitude is really what it is all about. People create most of their problems through their attitudes. Negative attitudes seem more contagious than positive ones.

You can create solutions to complicated problems by being the only one to break that complicated problem down to its basic causes.

A good follower has to want the same results the leader wants.

Many of the most frustrating and expensive problems seen today come from paperwork and similar communication devices.

What it comes down to is that the employees see their supervisor as "the company."

You need to lead people gently to what they already know is right.

If it is not already evident why these passages have been shared, let me tell you. The common threads that run through every different activity (safety, quality, marketing, manufacturing, labor relations, etc.) and how actions predict outcomes—good or bad—should edify us and not confound us. The commonality should help us find simpler solutions...not more complicated ones. VIRO will do that; one basic formula to solve lots of problems and help us predict and understand how to achieve more good outcomes and fewer bad ones.

Now to the enlightened business example. Managing a business successfully by values is not really new. The way this text will approach it is new, however. Has anyone talked about this before? You bet they have. On the cover of the August 1, 1994 issue of *BusinessWeek* is a picture of Bob Haas, the CEO of Levi Stauss. He is smiling and looks like "the cat that ate the canary." In big, black, block letters the title reads, "MANAGING BY VALUES." The magazine cover suggests that Haas is "putting his own strong views into practice." The issue promises to tell the reader how it is working and not working. Many companies that have highly developed cultures practice values that are well-known by most of the employees and could be detailed by many of them with noteworthy consistency. It's doubtful that many—if any—of them have called them company values or beliefs and then taught them systematically and measured the impact on the primary objectives of the organization.

BusinessWeek says that some companies are approaching diversity and empowerment as competitive tools and infer that this shift may be leading to an interest in managing by a set of values. Then they say, "No company, however, has embraced a values-based strategy the way Levi's has." Let's take a look at some of the things Haas was quoted in the article as saying:

"We started to improve at Levi's, when we stopped talking about values like diversity and started behaving that way."

"We are not doing this because it makes us feel good— although it does. We are not doing this because it is politically correct. We are doing this because we believe in the interconnection between liberating the talents of our people and business success."

"We are only a few steps along in our journey. We are far from perfect. We are far from where we want to be. But the goal is out there, and it's worth striving for."

"This is not a matter of structure, it is a matter of leadership." This was offered by Haas in answer to the suggestion that he wouldn't be doing this if he did not have controlling interest in the ownership of the company.

"People are comfortable with the traditional ways of doing business."

The focus on managing by values at Levi's occurred in a setting of significant corporate success until about six months before the article, when growth of sales and earnings started to flatten out. The article indicates that not everyone at Levi's supports Haas, including some board members. Anderson Consulting is quoted as suggesting that the process took longer and involved more people than they thought wise. Haas takes the long view and counters by saying that orienting the organization toward common goals through values will produce efficiency in the long run. That's refreshing and different, isn't it?

Haas is quoted as reflecting on the fact that the aspiration statements or statement of values were crafted by top management, not the human resources department. Apparently, the statements are printed on paper made from recycled denim and posted on office and factory walls throughout Levi's. This may not be a good idea, or at least the best idea.

What's missing by this account? There is no talk in the article of involving employees in determining what the values should be. *If that isn't done, it's a mistake.* There is no talk of measuring the penetration of the values in the organization or correlating them with improved performance. *If that isn't done, there is an important void in the process.* No direct linkage is detailed between the values and the vitality of the organization, except the statement about liberating people. *If that isn't done, an opportunity is being missed.* There is no talk in the article of extending the concept of managing by values down into individual departments and disciplines. *If that isn't being done, the foundation for the corporate activities won't be strong, and employees will have more trouble relating to the concepts and seeing the big picture.*

Levi Strauss appears to be at the cutting edge of where successful businesses will be operating in the next decades. I wish them well and hope they will stick to it in spite of criticism and naysayers. I have already outlined a second book that explores values-driven business management, i.e., how to use a foundation of values to make the whole enterprise successful, not just their loss prevention efforts. This book intends to aid Bob Haas and others like him. Maybe some of them can help me write the book.

A week after the Haas article last year, in the August 8, 1994 issue of *U.S. News & World Report,* the editor-in-chief, Mortimer B. Zuckerman wrote his editorial on values, "Where Have Our Values Gone?" Mr. Zuckerman starts by reflecting on the social dysfunction that "haunts the land" in the form of crime, chemical dependency, the slump in academics, and the break-up of the family. He effectively punctuates his introduction with a phrase quoted from Senator Daniel Patrick Moynihan: "Are we now, 'defining deviancy down,' accepting as part of life what we once found repugnant?"

Mr. Zuckerman points out that the habits America once admired, industriousness, thrift, self-discipline, and commitment, seem to be disappearing.

He sees them replaced by excuse making. "Crime is sanctioned by the fact, real or imagined, that the criminal had an unhappy childhood."

The editorial suggests that the solutions probably lie in education, not only as an economic imperative but as a moral imperative. The editor sees "a great yearning in the country to provide our national life and institutions with a larger, moral dimension" and offers the success of the movie *Forrest Gump* as testimony to that. In the end, he suggests that we have two choices: arrest the decay or allow the dysfunction to continue. He sees anxiety turning to fear. . . and then panic. When fear dominates social policy, he sees reason and tolerance being at risk. He's right.

The editorial is a call to action without a prescription for the action other than a suggestion for better and more complete education. The word *values* is only used once in the text. There is little linkage between beliefs, values, culture, and behaviors. Mr. Zuckerman got the ball rolling, but who's picking it up and running with it? The philosophy exposed earlier in this book may be the answer...if tailored to a wider subject: the moral and spiritual health of this nation.

I have also planned a third book to be titled *"Solutions;...To all the Problems (Troubles) in the World: The Ultimate How-To Book."* This bold title sounds—and is—presumptuous. And I confess that assuming to write a book offering a blueprint to solving national health, education, debt, and crime problems while accelerating personal development of the readers *is* lofty. But so is the nature and power of values.

EPILOGUE

The strength of our nation has always been our willingness—no, our desire—to sacrifice so our children and grandchildren could have a better life. Our healthiest enterprises have served as an extended family and have modeled the same philosophy, which has been to work hard so we will leave something good for future members/employees. Now we seem to have lapsed into being more concerned with our retirement benefits than with the quality of life we have bequeathed to our offspring. We have little interest in performing as a part of an extended family except for our own gratification or paycheck.

This isn't working, and it can never work. When the government runs out of money, will those children be eager to take us in when they realize what we have done *to* them instead of *for* them? If we don't give our all to the organization for which we work, will it still be there when we need it or will some foreign source be controlling our destiny? Remember the company that didn't care about their employees; should they be surprised when their employees aren't loyal and leave for greener pastures at a critical moment?

Have we changed our basic instincts? Have we suddenly become selfish? I don't think so. I just think we lost our way because we forgot what got us here. It didn't happen over night, so, we won't find our way back over night. And, we will need a map, a good guide, and some light. We'll need to recognize our destination when we get there. I think VIRO and what can grow from it could help guide us. This is not just a safety program. It is a whole new way to look at the world around us. I am suggesting that we pilot it on the subject of loss prevention, the optimization of our physical and human resources.

In the Prologue, I detailed what I intended to accomplish in the book. I promised to review those objectives and how I did against them in the Epilogue.

First, I promised to substantiate six premises that were linked together and which formed the core philosophy on which the book is based. The premises are: 1) We are governed by social laws that are just as irrevocable as the physical laws that govern us. 2) Solving problems depends on accurate diagnosis. 3) Beliefs and values predict outcomes. 4) The methods used to minimize loss will work just as well to optimize desired outcomes in quality or in individual lives. 5) Beliefs and values can be influenced and taught. 6) Shaping outcomes at the source or inception of the activity is far easier and more effective than late in the process. *I shared many stories, examples, anecdotes, and analogies. These, I hope, have convinced you that my premises are essentially accurate. If VIRO is to work for you, you need to subscribe to these six premises.*

Second, I said I'd suggest values that predict freedom from undesired outcomes. *I list and examine ten of those in Section II and even suggest alternates or supplements.*

Third, I promised to explain and justify the values. *I think I did that in detail. . . perhaps too much detail at times.*

Fourth, I indicated I would provide a way to measure values. *The maturity grids and suggested methods for use described in Chapter 18 constitute the promised measurement system.*

Fifth, I suggested I would detail exactly how an organization could apply the concept. *This is a little fuzzy. I do it within the discussion of each value and in the suggested methods to install and measure. I am anxious to hold seminars on this subject and work with groups of enlightened people to brainstorm how this can be done. My ideas, I believe, are just the beginning. Look how long it took us to realize all the things we could do with electricity. And we're still learning. From a social standpoint, I see the VIRO concept as having the same kind of potential. If I had done this in too much detail, I would not have been able to avoid culture blindness.*

Sixth, I counseled on the need to correlate VIRO measurement with the more traditional techniques in use and said I would provide ideas on how that could be done. *I have provided ideas, but this exercise will be highly individualized. You will need to take my thoughts plus some of your own and apply them to your individual situation.*

Seventh, and lastly, I said I would convince you that VIRO can and will work if applied correctly. *You probably noticed that I spent much of the book trying to persuade readers of the validity and power of the concept of driving safety efforts through organizational values. The evidence is overwhelming.*

I hope you have enjoyed the book and will apply its teachings. Feel free to contact me with your questions, feedback, or ideas. I'm excited about extending the concept of using values based on correct principles to achieve benefits beyond the traditional safety realm. I hope to have at least one more book written on this subject before you read this one.

BIBLIOGRAPHY

Beard, Henry, & Cerf, Christopher. *The Official Politically Correct Dictionary and Handbook*. New York: Villard Books, 1992.

Bennett, William J. *The Book of Virtues*. New York: Simon & Schuster, 1993.

Business Week. "Managing by Values," August 1, 1994.

Carey, Art. *The United States of Incompetence*. Boston, MA: Houghton Mifflin Company, 1991.

Chopra, Deepak. *The Seven Spiritual Laws of Success*. San Rafael, CA: Amber-Allen Publishing, 1994.

Collins, James C., & Porras, Jerry I. *Built to Last*. New York: HarperBusiness, 1994.

Committee on Trauma Research, Commission on Life Sciences, National Research Council, the Institute of Medicine. *Injury in America*. Washington, D.C.: National Academy Press, 1985.

Covey, Stephen R. *First Things First*. New York: Fireside/Simon & Schuster, 1994.

Covey, Stephen R. *Principle-Centered Leadership*. New York: Fireside/Simon & Schuster, 1990.

Covey, Stephen R. *The 7 Habits of Highly Effective People*. New York: Fireside/Simon & Schuster, 1989.

Crosby, Philip B. *Quality Is Free*. New York: McGraw-Hill Book Company, 1979.

Culbertson, Charles V. *Managing Your Safety Manager*. New York: Risk and Insurance Management Society, Inc., 1981.

Drucker, Peter F. *The New Realities*. New York: Harper & Row, Publishers, 1989.

Drucker, Peter F. *Management Tasks Responsibilities Practices*. New York: Harper & Row, Publishers,1973.

Eadie, Betty J. *Embraced by the Light*. Placerville, CA: Gold Leaf Press, 1992.

Eckenfelder, Donald J. "A Ten-Step Strategy for Loss Prevention," *Risk Management*, May 1992.

Eckenfelder, Donald J. "Safety Plans Are Key to Cutting Workers Comp Costs," *Human Resources Professional*, Fall 1991.

Eckenfelder, Donald J., & Zaledonis, Charles E. "Engineering project planner, a way to enfineer out unsafe conditions," *Professional Safety*, November 1976.

Eyre, Linda and Richard. *Teaching Your Children Values*. New York: Fireside/Simon & Schuster, 1993.

Eyre, Richard, *Don't Just Do Something, Sit There*. New York: Fireside/Simon & Schuster, 1995.

Gingrich, Newt, *To Renew America*. New York: Harper Collins, 1995.

Grimaldi, John V., & Simonds, Rollin H. *Safety Management*. Homewood, IL: Richard D. Irwin, Inc., 1956.

Hammer, Michael, & Champy, James. *Reengineering the Corporation*. New York: HarperBusiness, 1993.

Hansen, Larry. "Safety Management: A Call for (R)evolution," *Professional Safety*, March, 1993.

Hostage, G. M. "The line manager and his safety professional—how to prevent accidents," *Professional Safety*, November, 1996.

Howard, Philip K. *The Death Of Common Sense*. New York: Random House, 1994.

Huffington, Arianna. *The Fourth Instinct*. New York: Simon & Schuster, 1994.

Krause, Thomas R., & Hidley, John H., & Hodson, Stanley J. *The Behavior-Based Safety Process*. New York: Van Nostrand Reinhold, 1990.

Lareau, William. *American Samurai*. Clinton, NJ: New Win Publishing, Inc., 1991.

Lowrance, William W. *Of Acceptable Risk*. Los Altos, CA: William Kaufmann, Inc., 1976.

Mieder, Wolfgang. *A Dictionary of American Proverbs*. New York: Oxford University Press, 1992.

Peters, Thomas J., & Waterman, Robert H. *In Search of Excellence*. New York: Harper & Row, 1982.

Peters, Tom. *Liberation Management*. New York: Alfred A. Knopf, 1992.

Peterson, Dan. *Safety Management: A Human Approach*. Englewood, New Jersey: Aloray Publisher, 1975.

Pierce, F. David. *Total Quality for Safety and Health Professionals*. Rockville, MD: Government Institutes, Inc., 1995.

Short, Robert L. *The Parables of Peanuts*. New York: Harper & Row, 1968.

Tarrants, William E. *The Measurement of Safety Performance*. New York: Garland Publishing, Inc., 1980.

Tarrant, John J. *Drucker, the Man Who Invented the Corporate Society*. New York: Warner Books, 1976.

Tobias, Andrew. *The Invisible Bankers*. New York: The Linden Press/Simon & Schuster, 1982.

Trimble, John R. *Writing with style*. Englewood Cliffs, NJ: Prentice-Hall, Inc., 1975.

Wattenberg, Ben J. *Values Matter Most*. New York: The Free Press, 1995.

Woodward, Bob. *The Agenda*. New York: Simon & Schuster, 1994.

Zuckerman, Mortimer B. "Where Have All Our Values Gone?" *U.S. News & World Report,* Editorial, August 8, 1994.

INDEX

Government Institutes Mini-Catalog

PC #	**ENVIRONMENTAL TITLES**	Pub Date	Price
629	ABCs of Environmental Regulation: Understanding the Fed Regs	1998	$49
627	ABCs of Environmental Science	1998	$39
585	Book of Lists for Regulated Hazardous Substances, 8th Edition	1997	$79
579	Brownfields Redevelopment	1998	$79
4088 ⬛	CFR Chemical Lists on CD ROM, 1997 Edition	1997	$125
4089 ⬛	Chemical Data for Workplace Sampling & Analysis, Single User Disk	1997	$125
512	Clean Water Handbook, 2nd Edition	1996	$89
581	EH&S Auditing Made Easy	1997	$79
587	E H & S CFR Training Requirements, 3rd Edition	1997	$89
4082 ⬛	EMMI-Envl Monitoring Methods Index for Windows-Network	1997	$537
4082 ⬛	EMMI-Envl Monitoring Methods Index for Windows-Single User	1997	$179
525	Environmental Audits, 7th Edition	1996	$79
548	Environmental Engineering and Science: An Introduction	1997	$79
643	Environmental Guide to the Internet, 4rd Edition	1998	$59
560	Environmental Law Handbook, 14th Edition	1997	$79
353	Environmental Regulatory Glossary, 6th Edition	1993	$79
625	Environmental Statutes, 1998 Edition	1998	$69
4098 ⬛	Environmental Statutes Book/CD-ROM, 1998 Edition	1997	$208
4994 ⬛	Environmental Statutes on Disk for Windows-Network	1997	$405
4994 ⬛	Environmental Statutes on Disk for Windows-Single User	1997	$139
570	Environmentalism at the Crossroads	1995	$39
536	ESAs Made Easy	1996	$59
515	Industrial Environmental Management: A Practical Approach	1996	$79
510	ISO 14000: Understanding Environmental Standards	1996	$69
551	ISO 14001: An Executive Report	1996	$55
588	International Environmental Auditing	1998	$149
518	Lead Regulation Handbook	1996	$79
478	Principles of EH&S Management	1995	$69
554	Property Rights: Understanding Government Takings	1997	$79
582	Recycling & Waste Mgmt Guide to the Internet	1997	$49
603	Superfund Manual, 6th Edition	1997	$115
566	TSCA Handbook, 3rd Edition	1997	$95
534	Wetland Mitigation: Mitigation Banking and Other Strategies	1997	$75

PC #	**SAFETY and HEALTH TITLES**	Pub Date	Price
547	Construction Safety Handbook	1996	$79
553	Cumulative Trauma Disorders	1997	$59
559	Forklift Safety	1997	$65
539	Fundamentals of Occupational Safety & Health	1996	$49
612	HAZWOPER Incident Command	1998	$59
535	Making Sense of OSHA Compliance	1997	$59
589	Managing Fatigue in Transportation, *ATA Conference*	1997	$75
558	PPE Made Easy	1998	$79
598	Project Mgmt for E H & S Professionals	1997	$59
552	Safety & Health in Agriculture, Forestry and Fisheries	1997	$125
613	Safety & Health on the Internet, 2nd Edition	1998	$49
597	Safety Is A People Business	1997	$49
463	Safety Made Easy	1995	$49
590	Your Company Safety and Health Manual	1997	$79

Government Institutes

4 Research Place, Suite 200 • Rockville, MD 20850-3226
Tel. (301) 921-2323 • FAX (301) 921-0264
Email: giinfo@govinst.com • Internet: http://www.govinst.com

Please call our customer service department at
(301) 921-2323 for a free publications catalog.

CFRs now available online.
Call (301) 921-2355 for info.

GOVERNMENT INSTITUTES ORDER FORM

4 Research Place, Suite 200 • Rockville, MD 20850-3226
Tel (301) 921-2323 • Fax (301) 921-0264
Internet: http://www.govinst.com • E-mail: giinfo@govinst.com

3 EASY WAYS TO ORDER

1. Phone: **(301) 921-2323**
Have your credit card ready when you call.

2. Fax: **(301) 921-0264**
Fax this completed order form with your company purchase order or credit card information.

3. Mail: **Government Institutes**
4 Research Place, Suite 200
Rockville, MD 20850-3226 USA
Mail this completed order form with a check, company purchase order, or credit card information.

PAYMENT OPTIONS

❏ **Check** *(payable to Government Institutes in US dollars)*

❏ **Purchase Order** *(This order form must be attached to your company P.O. <u>Note</u>: All International orders must be prepaid.)*

❏ **Credit Card** ❏ VISA ❏ ⬤⬤⬤ ❏ ▭▭▭

Exp.___/____

Credit Card No. _____

Signature _____

(Government Institutes' Federal I.D.# is 13-2695912)

CUSTOMER INFORMATION

Ship To: (Please attach your purchase order)

Name: _____

GI Account # *(7 digits on mailing label):* _____

Company/Institution: _____

Address: _____
<u>(Please supply street address for UPS shipping)</u>
_____

City: _____ State/Province: _____

Zip/Postal Code: _____ Country: _____

Tel: ( ) _____

Fax: ( ) _____

Email Address: _____

Bill To: (if different from ship-to address)

Name: _____

Title/Position: _____

Company/Institution: _____

Address: _____
<u>(Please supply street address for UPS shipping)</u>
_____

City: _____ State/Province: _____

Zip/Postal Code: _____ Country: _____

Tel: ( ) _____

Fax: ( ) _____

Email Address: _____

Qty.	Product Code	Title	Price

❏ **New Edition No Obligation Standing Order Program**
Please enroll me in this program for the products I have ordered. Government Institutes will notify me of new editions by sending me an invoice. I understand that there is no obligation to purchase the product. This invoice is simply my reminder that a new edition has been released.

15 DAY MONEY-BACK GUARANTEE
If you're not completely satisfied with any product, return it undamaged within 15 days for a full and immediate refund on the price of the product.

Now Order Online: www.govinst.com

Subtotal _____
MD Residents add 5% Sales Tax _____
Shipping and Handling (see box below) _____
Total Payment Enclosed _____

Within U.S:	**Outside U.S:**
1-4 products: $6/product	Add $15 for each item (Airmail)
5 or more: $3/product	Add $10 for each item (Surface)

SOURCE CODE: BP01